Vulnerable Systems

Wolfgang Kröger · Enrico Zio

with contributions of
Markus Schläpfer · Cen Nan · Konstantinos Trantopoulos
Giovanni Sansavini · Miltos Kyriakidis

Vulnerable Systems

Springer

Prof. Dr. Wolfgang Kröger
Mechanical and Process Engineering Department
ETH Zurich
Sonneggstrasse 3
8092 Zurich
Switzerland
e-mail: kroeger@mavt.ethz.ch

Prof. Dr. Enrico Zio
Laboratoire Génie Industriel
Ecole Central Paris and Supelec
Grande Voie des Vignes
92295 Chatenay-Malabry Cedex
France
e-mail: enrico.zio@ecp.fr, enrico.zio@supelec.fr

Prof. Dr. Enrico Zio
Dipartimento di Energia
Politecnico di Milano
Via Ponzio 34/3
20133 Milano
Italy
e-mail: enrico.zio@polimi.it

Markus Schläpfer
Cen Nan
Konstantinos Trantopoulos
Swiss Federal Institute of Technology
Zürich (ETH)
Switzerland

Giovanni Sansavini
Polytechnic of Milan
Milan
Italy

Miltos Kyriakidis
Imperial College
London
UK

ISBN 978-0-85729-654-2 e-ISBN 978-0-85729-655-9
DOI 10.1007/978-0-85729-655-9
Springer London Dordrecht Heidelberg New York

British Library Cataloguing in Publication Data
A catalogue record for this book is available from the British Library

Cover design: eStudio Calamar, Berlin/Figueres

Printed on acid-free paper

Springer is part of Springer Science+Business Media (www.springer.com)

Preface

The attention on critical infrastructures is evolving from concerns about aging public works (in the 1980s) to redefinition in terms of national security as a result of increased international terrorism (after 11 September 2001) and susceptibility against natural hazards, to unprecedented failure combinations and malicious (cyber) attacks (during mid-2000s). Consequently, the view has broadened from local via national/regional to global while concerns about single failure mechanisms have developed into a full set of potential failures, hazards and threats. As a result, nowadays strategies to reduce and manage vulnerabilities in critical infrastructures and the provision of related analytical instruments have to follow an "all hazards approach".

Societies have always been dependent on services provided by infrastructures, but recently the systems involved have witnessed higher and tighter integration, e.g., by means of pervasive use of modern information and communication technology (ICT), changing operational environments, e.g., market liberalization, and growing mutual dependencies. At the same time the public has become even more dependent on the continuous service infrastructures offer (the Internet and consumption of electricity as a common good may serve as examples). Then today, we cannot allow many such systems to debilitate or collapse, as inconveniences and risks are unacceptable and financial losses are huge, e.g., the 14 August 2003 blackout in North America affected 50 million people and led to 3 billion USD insurance claims.

From the analysis point of view, it is commonly agreed that infrastructures have become more complex and their behavior is hard to understand or predict; research on complex networks has shown that some elements (nodes) evolve to become more important and some structures (topologies) are more susceptible against random failures or targeted attacks than others. Reduction of technological and social vulnerabilities calls for better system understanding and preventive analyses, efforts for which established frameworks and methods seem to be still missing or, if available, are not properly applied.

Infrastructures are of different types and dimensions and show different importance for our society, i.e. different degrees of criticality. This book will focus on large, geographically-distributed, physical-engineered networks, and particularly on those of undoubted importance for at least highly industrialized countries, such as:

- energy supply (electricity, gas)
- urban freshwater supply and wastewater treatment
- information and communication
- transport (rail, road)
- control systems (SCADA).

Most of these systems are coupled and mutually dependent on different degree and order, difficult to understand and emulate. Very often "the system" consists of a "part being under control" and "the control part" itself, using the same technology as public information and communication systems or even those directly, i.e., the Internet for transport of data and commands. This book will reflect these structural conditions, where appropriate, when elaborating challenges to methods and assessing their quality/suitability.

The aim of the book is of collecting and capturing the state of knowledge in the evolving research field of vulnerability analysis of complex physical-engineered networks, by literature survey focused on "landmark" reports and making use of ongoing authors' work. The targeted audience encompasses students of natural and engineering sciences at MS or PhD level, researchers in the field, non-routine practitioners in industries and agencies as well as executives responsible for critical infrastructure design and protection.

The book is structured according to the course of a fictitious vulnerability analysis. At first the book introduces critical infrastructures in a top-down manner and defines the key terms including the concept of vulnerability followed by elaborations on characteristics of critical infrastructure such as complexity and dimensions of interdependencies, and on challenges to methods. Approaches for vulnerability assessment are categorized and outlined in more general terms before—in the focal chapter—some methods for screening analysis and in-depth analysis are presented in detail. The methods are selected according to their eligibility to meet the challenges posed by the basic characteristics of the types of infrastructures and the objectives of the analysis defined beforehand. Selected methods, e.g. complex network theory, probabilistic techniques, and object-oriented modeling, will be introduced in terms of basic approach, algorithm and measures, and explained by means of illustrative examples; strengths and weaknesses will be assessed separately and finally comparatively. Well-established and well-known methods, e.g., Petri nets, will be mentioned but not explained in detail; reference books will be referred to.

Acknowledgements

The authors wish to express their sincere gratitude to all those who have contributed to this book by participating in basic discussions, contributing to chapters and reviewing drafts, namely Giovanni Sansavini (basic discussions and contributions to Chap. 4, Sect. 6.2, 6.3 and 6.4), Irene Eusgeld (basic discussions, input to Chap. 3), Adrian Gheorghe (review Sect. 6.6), Lucia Roxana Golea (contribution to Sect. 6.3), Arnab Majumdar and Vinh Dang (supervision and review of Sect. 6.7), and Roberta Piccinelli (contribution to Sect. 6.4).

The authors would also like to express their high appreciation to Sabine Keller for typewriting of chapters and, in particular for providing Word files and Patrick Probst for general support and providing few image files.

Contents

Abbreviations

ABM	Agent-based modeling
ALMs	Accelerated lifetime models
ALSP	Aggregate level simulation protocol
ATHEANA	A technique for human error analysis
BNs	Bayesian networks
CARA	Controller action reliability assessment
CCDF	Complementary-cumulative distribution function
CDM	Communication data and management
CFP	Cognitive failure probability
CIs	Critical infrastructures
CREAM	Cognitive reliability and error analysis method
CTMC	Continuous-time Markov chain
DAWG	Data analysis working group
DISD	Distributed interactive simulation
DMSO	U.S. Defense modeling and simulation office
DOE	U.S. Department of energy
EHV	Extra-high voltage
ENTSO-E	European network of transmission system operators for electricity
EPOCHS	Electric power and communication synchronizing simulator
EPS	Electric power system
EPSS	Electric power supply system
ET	Event tree
FLSC	Swedish air force air combat simulation centre
FOM	Federate object model
FSM	Finite state machine
FMEA	Failure modes and effects analysis
FT	Fault tree
GLMs	Generalized linear models
GVW	Geographic valued worth
HAZOP	Hazard and operability

HEART	Human error assessment and reduction technique
HLA	High level architecture
HRA	Human reliability analysis
ICT	Information communication technology
IEEE	Institute of electrical and electronic engineers
IEs	Initiating events
IRRIS	Integrated risk reduction of information-based infrastructure systems
ISS	Interactive simulation systems
LAN	Local area network
LTI	Linear time invariant
MCs	Markov chains
MAUT	Multi-attitude utility theory
MCS	Minimal cut set
MLDs	Master logic diagrams
MPNs	Markov/Petri nets
MTTF	Mean-time-to-failure
NERC	North american electric reliability council
NET	Network event tree
NSF	National science foundation
OMT	Object model template
OPF	Optimal power flow
PHMs	Proportional hazard models
PI	Performance index
PRA	Probabilistic risk assessment
QRA	Quantitative risk assessment
RTI	Run time infrastructure
RTU	Remote terminal unit
SCADA	Supervisory control and data acquisition
SHERPA	Systematic human error reduction and prediction approach
SLIM/MAUD	Success likelihood index method/multi-attribute utility decomposition
SOC	Self-organized criticality
SOM	Simulate object model
SUB	Substation
THERP	Technique for human error rate prediction
TSO	Transmission systems operator
UCTE	Union for the coordination of transmission of electricity
UML	Unified modeling language
WAN	Wide area network

Chapter 1
Introduction and Definition of Key Terms

The welfare and security of each nation rely on the continuous flow of essential goods (such as energy and data) and services (such as banking and health care). A large-scale array of wide area, man-made systems and assets, mostly privately owned or operated, that function collaboratively and synergistically to produce and distribute such a flow, are called infrastructures. Those infrastructures so vital to any country that their incapacity or destruction would have a debilitating impact on the health, safety, security, economics, and social well-being, including the effective functioning of governments[1] are called critical. A failure within one of these infrastructures, or the loss of its continuous service, may be damaging enough to a society and its economy, while that which cascades across boundaries has the potential for multi-infrastructural collapse and unprecedented consequences.

Critical infrastructures (CIs) are various by nature, e.g., physical-engineered, cybernetic or organizational systems, and by environment (geographical, natural) and operational context (political/legal/institutional, economic, etc.).

In principle, a system can be defined as a group of interacting elements (or subsystems) having an internal structure and comprising a unified whole. The boundary of a system is either given or obvious or needs to be defined. Autonomy, coherence, permanence, and organization are essential properties of systems (Dupuy 1985).

Engineered physically networked CIs, often called lifeline systems, is the focus of this book; examples are those providing:

- Energy (electricity, oil, and gas supply)
- Transportation (by rail, road, air, and sea)
- Information and telecommunication (such as the Internet)
- Drinking water, including wastewater treatment

[1] Definition refers to President's Commission on Critical Infrastructure Protection (1997), USA Patriot Act (2001) and European Commission (2004) but was slightly modified by the authors.

W. Kröger and E. Zio, *Vulnerable Systems*, DOI: 10.1007/978-0-85729-655-9_1,

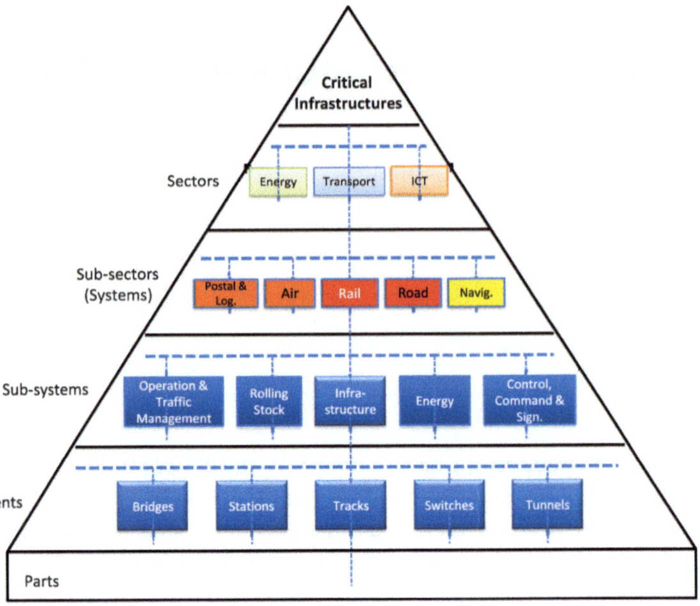

Fig. 1.1 Hierarchical representation of the rail system

The system of CIs can be represented by hierarchical layers which are linked through physical and logical relations (see Fig. 1.1 for the rail transport system).

These CIs are subject to a set of multiple hazards and potentially asymmetrical threats (Table 1.1) disclosing weaknesses and vulnerabilities, respectively; furthermore, they may pose risks themselves during normal operation (e.g., electromagnetic fields—EMF) or accidents (e.g., rupture of gas pipelines). Also, most CIs have a dynamic structure, are undergoing far-reaching changes, both technological and organizational, and incorporating technologies soon after they are (commercially) available.

As shown by experienced events, CIs are highly interconnected and mutually dependent in complex ways, both physically and through a host of information and communication technologies, the so-called cyber-based systems (Rinaldi et al. 2001). Identifying, understanding and analyzing these features are still major challenges, magnified by the breadth and complexity[2] of most infrastructures.

According to Rinaldi et al. (2001), *dependency* is defined as a unidirectional relationship between two infrastructures, that is infrastructure i depends on j through the link, but j does not depend on i through the same link, while *interdependency* defines a bidirectional relationship, that is infrastructure i depends on j through some links, and j likewise depends on i through the same and/or other links.

---

[2] See Chap. 2 for definition.

Table 1.1 Set of multiple hazards and threats disclosing vulnerabilities of CI

Natural events such as earthquakes, hurricanes/typhoons, tornados, severe flooding, landslides or other (increasingly) extreme weather conditions
Accidents or technical factors such as components' failure/rupture leading to the debilitation of plants, networks and operations
Market factors such as instability associated with major producer groups, or economic pressure trading off security factors
Policy factors such as artificial supply limitations or negative pricing outcomes or misusing "energy" for political purposes
Human factors such as unintended failures of omission or commission, e.g., of system operator, intended errors or even targeted malicious attacks, either physical or cyber

Infrastructures are not only complex but most of them show adaptive behavior; that is, all components and the system as a whole are influenced by past experience, e.g., degradation from overuse, aging over time, by trials to improve performance, e.g., of the personnel, and by adjustment to new conditions or disturbances, e.g., automatic variation of generator output to meet actual power loads or load-shedding (Rinaldi et al. 2001).

Infrastructure interdependencies[3] have often been illustrated by a dependency matrix (IRGC 2005; Luiijf 2008) or by representing infrastructure networks as interconnected single planes or layers as shown in Fig. 1.2. In Fig. 1.2, parallel lines represent individual sectors or subjects within a particular infrastructure; solid lines connect nodes and cross-sectors in internal dependencies while dashed lines mark interdependencies. A meaningful representation of such web of dependencies must relate to a specific scenario; here, the flooding event and subsequent response during Hurricane Katarina have been selected.

The *vulnerability* of CIs must be theoretically analyzed and assessed. While the concept of risk is fairly mature and consensually agreed, the *concept of vulnerability* is still evolving and not yet established. In general terms, *risk* refers to a combination of the probability of occurrence (frequency *F*) of a specific (mostly undesired/adverse) event leading to loss, damage or injury and its extent (consequence indicators c_j)[4]. These quantities and their associated uncertainties are regarded as being numerically quantifiable. Besides this quantitative side of risk, there is a non-technical dimension accounting for the aspects of societal and psychological risk experience and perception which are subject to changes and contextual in nature.[5] For CIs, the term risk may include the frequency of loss of service with its resulting consequences for the people concerned.

[3] See Chap. 2 for further explanation.

[4] See also ISO/IEC Guide 73 (ISO/IEC 2002).

[5] See also German Advisory Council for Global Change (WBGU 1999) and IRGC White Paper 1 (IRGC 2005, p 141) for further details.

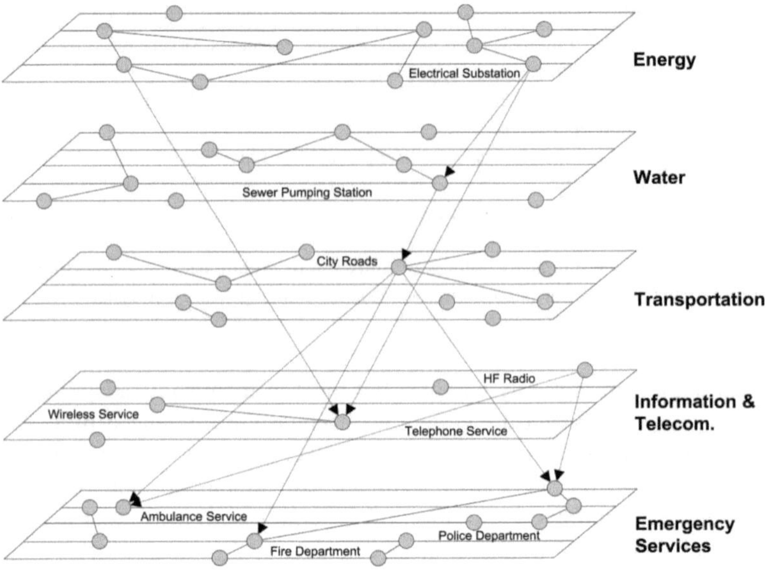

Fig. 1.2 Infrastructure interdependencies, illustrated for the flooding event and subsequent response during Hurricane Katrina, USA (Pederson et al. 2006)

The term *vulnerability* has been introduced as the hazard[6]-centric perception of disasters that is revealed as being too limited to understand in terms of risks. A hazard of low intensity could have severe consequences, while a hazard of high intensity could have negligible consequences: the level of *vulnerability* is making the difference (White 1974).

The *concept of vulnerability* seen as a global system property focuses on three elements[7]:

- Degree of loss and damages due to the impact of a hazard (technical dimension)
- Degree of exposure to the hazards, i.e., likelihood of being exposed to hazards of a certain degree and susceptibility of an element at the risk of suffering loss and damages (the element at risk could be a technical system)
- Degree of resilience,[8] i.e., the ability of a system to anticipate, cope with/absorb, resist and recover from the impact of a hazard (technical) or disaster (social).

[6] "A potentially damaging physical event, phenomenon and/or human activity, which may cause loss of life or injury, property damage, social and economic disruption or environmental degradation. Hazards can be single, sequential or combined in their origin and effects" (UN/ISDR 2004).

[7] See Bouchon (2006) for further detailed explanations.

[8] Resilience generally means the ability to recover from some shock, insult, or disturbance, the quality or state of being flexible. In *physics and engineering*, it is defined as the physical property of a material that can return to its original shape or position after deformation that does not exceed its elastic limit, i.e., as its capacity to absorb energy when it is deformed and then, upon

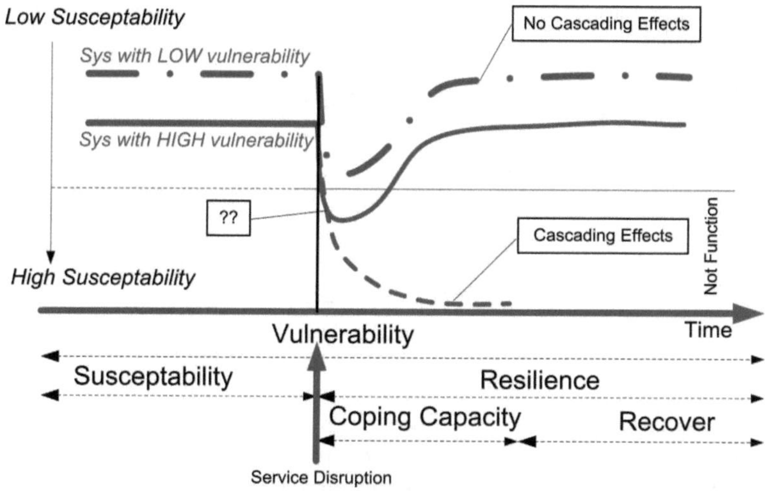

Fig. 1.3 Vulnerability elements and associated response scenarios (Bouchon 2006)

Figure 1.3 brings these elements together with the scenarios which may develop depending on the system characteristics; cascading effects are shown to possibly lead to a complete system breakdown.

In the context of the material presented in this book, we define *vulnerability* as a flaw or weakness (inherent characteristic, including resilience capacity) in the design, implementation, operation, and/or management of an infrastructure system, or its elements, that renders it susceptible to destruction or incapacitation when exposed to a hazard or threat, or reduces its capacity to resume new stable conditions. The latter can be provided with a likelihood (frequency) while a measurand for destruction or incapacitation (loss or damage, respectively) needs specific elaborations depending on the value placed on the asset by its owner/operator or the customer/government. For example, the *vulnerability* of the electric power system might be specified in terms of changes in network characteristics following attacks on nodes and the scale (e.g., number of nodes/lines lost) or the duration of the associated loss. More sophistically, it can be expressed in terms of the frequency of major blackouts (number per year) and the associated severity, measured either in power lost or energy unserved (MW or MWh) as illustrated by Fig. 1.4.

Therefore, this interpretation of *vulnerability* is closely related to the definition of risk while another interpretation is used to describe a system component or an aspect of a system, i.e., a component is said to be a *vulnerability* of a system if its

(Footnote 8 continued)
unloading, to have this energy recovered. Regarding systems resilience basically it is the potential to remain in a particular configuration and to maintain its feedback and functions, and involves the ability of the system to reorganize following disturbance-driven changes (Bouchon 2006).

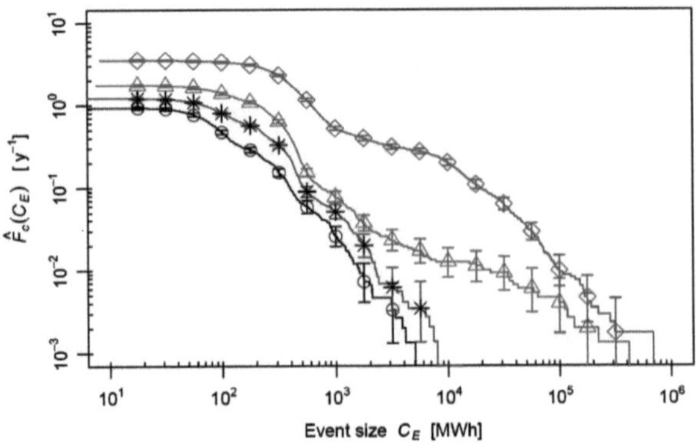

Fig. 1.4 Complementary cumulative blackout frequencies for four different grid load levels L-100% (*circles*), 110% (*stars*), 120% (*triangles*) and 137% (*diamonds*) (Schläpfer et al. 2008)

failure causes large negative consequences to that system (Jönsson et al. 2008). The measure could be a ranking of components a system depends upon.

Reliability of an infrastructure of interest and availability of a service or goods it provides are also attributes useful to subscribe the quality of infrastructure systems. These terms are defined as follows:

- *Reliability* is the ability of a system or component to perform its required functions under stated conditions for a specified period of time (mission without maintenance). Reliability is quantitatively expressed as a probability.
- *Availability* is the probability of a unit to be in working state at a given time t (includes maintenance).

The term "safety" is defined as the absence of a specified damage on users, the public and the environment taking unintentional (random) triggering acts/events, failures or faults into account while "security" includes threats of intentional origin such as sabotage, cyber attacks and terrorism. The traditional security attributes such as availability, confidentiality and integrity are often applied, together with attributes such as privacy and accountability (see also Aven 2007).

Given the realm of single infrastructures and interdependencies, the*goals of vulnerability analysis*, and the associated modeling and simulation efforts, could be:

1. Given a system and the end state of interest, identify the set of events and event sequences that can cause damages and loss effects.
2. Identify the relevant set of "initiating events" and evaluate their cascading impact on a subset of elements, or the system as a whole.
3. Given a system and the end state of interest, identify the set of events or respective event sequences that would cause this effect.
4. Given the set of initiating events and observed outcomes, determine and elaborate on (inter)dependencies (within the system and among systems) and on coupling of different orders.

The ultimate goal is to identify obvious and, most importantly, hidden vulnerabilities in infrastructure systems, to be able to act for managing and reducing them. The achievement of these goals rely on the analysis of the system, its parts and their interactions within the system; the analysis must account for the environment which the system lives in and operates, and finally for the objectives the system is expected to achieve. During the development of such basic system understanding, first vulnerabilities may have already been emerged.

References

Aven T (2007) A unified framework for risk and vulnerability analysis covering both safety and security. Reliab Eng Syst Safe 92(6):745–754

Bouchon S (2006) The vulnerability of interdependent critical infrastructures systems: epistemological and conceptual state-of-the-art (EU report). EU Commission, Joint Research Centre, Ispra, Italy

Dupuy G (1985) Systèmes, réseaux et territoires. Presse de L'Ecole nationale des Ponts et Chaussées, Paris

European Commission (2004) Critical infrastructure protection in the fight against terrorism, COM (2004) 702 final. Bruxelles, 20 October 2004

IRGC (2005) Risk governance: towards an integrative approach. White Paper No. 1, written by O Renn with an annex by P Graham. Geneva

ISO/IEC (2002) ISO/IEC Guide 73: risk management: vocabulary: guidelines for use in standards. ISO Technical Management Board Working Group 2

Pederson P, Dudenhoeffer D, Hartley S, Permann M (2006) Critical infrastructure interdependency modeling: a survey of U.S. and international research. Technical report INL/EXT-06-11464. Idaho National Laboratory, Idaho, USA

President's Commission on Critical Infrastructure Protection (1997) Critical foundations: protecting America's infrastructures. The report of the President's Commission on Critical Infrastructure Protection, Washington, DC

Rinaldi SM, Peerenboom JP, Kelly TK (2001) Identifying, understanding, and analyzing critical infrastructure interdependencies. IEEE Contr Syst Mag 21(6):11–25

Schläpfer M, Kessler T, Kröger W (2008) Reliability analysis of electric power systems using an object-oriented hybrid modeling approach. In: Proceedings of the 16th power systems computation conference, Glasgow, 14–18 July 2008

UN/ISDR (2004) Terminology: basic terms of disaster risk reduction: glossary. UN/ISDR, New York, USA

USA Patriot Act (2001) Uniting and strengthening America by providing appropriate tools required to intercept and obstruct terrorism (USA Patriot Act) act of 2001. H.R. 3162, in the Senate of the United States

White GF (1974) Natural hazards: local, national and global. Oxford University Press, New York p 288

Wissenschaftlicher Beirat der Bundesregierung Globale Umweltveränderungen (1999) Welt im Wandel: strategien zur Bewältigung Globaler Umweltrisiken/WBGU. Springer, Berlin, Heidelberg

Chapter 2
Properties of Critical Infrastructures

Physical-engineered critical infrastructures (CIs) are characterized as large scale, spatially distributed, complex networks—either open or closed. According to Dueñas-Osorio and Vemuru (2009), these systems are made of "a large number of interacting components (real or virtual), show emergent properties difficult to anticipate from the knowledge of single components, are characterized by a large degree of adaptability to absorb random disruptions and are highly vulnerable to widespread failure under adverse conditions." Indeed, small perturbations can trigger cascades and large-scale consequences in CIs; furthermore, disruptions may also be caused by targeted malicious attacks.

2.1 Complexity

A recent National Science Foundation (NSF) workshop report (Guckenheimer and Ottino 2008) points at the fact that a complex system is characterized by an internal structure which may consist, besides many interacting components, of "a network that describes which components of the system interact, multiple scales of space and/or time, or symmetry. The components of many complex systems are heterogeneous and form a hierarchy of subsystems." Furthermore, uncertainty is regarded as pervasive in complex systems, and its characterization and propagation through the system as key aspects for the reliable prediction of the system behavior and its effect and safe control.

The above attributes draw the boundary between simple and complex systems. Less trivial is to draw a boundary between complicated and complex systems. Table 2.1 attempts to do so by highlighting the very essence of a complex system, which is believed to lie in the degree and modality the parts interact and the overall behavior of the system that emerges from these. "The system must be analyzed as a whole; decomposing the system and analyzing subsystems does not necessarily give a clue as to the behavior of the whole" (Guckenheimer and Ottino 2008).

W. Kröger and E. Zio, *Vulnerable Systems*, DOI: 10.1007/978-0-85729-655-9_2,

Table 2.1 Characteristics of complicated versus complex systems, both entailing a large number of highly connected components

Complicated systems (mechanical watches, aircraft, power plants, etc.)	Complex systems (stock market, www, power grid, etc.)
Components have well-defined roles and are governed by prescribed interactions	Rules of interaction between the components may change over time and may not be well understood
Structure remains stable over the time. Low dynamics	Connectivity of the components may be quite plastic and roles may be fluid. Interactions are not always obvious
No adaptation. One key defect may bring system to a halt	System responds to external conditions and evolves
Limited range of responses to changes in their environment	Display organization without a central organizing principle (self-organization/emergence)
Decomposing the system and analyzing sub-parts can give us an understanding of the behavior of the whole, i.e., the whole can be reassembled from its parts	Respond to and interact with their environment
Problems can be solved through analytical thinking and diligence work	Inadequate information about the state of the influencing variables, nonlinearities
	Overall behavior cannot be simplified in terms of their building blocks. The whole is much more than the sum of its parts

2.2 Learning from Experience

Despite Cassandra, CIs have proved highly reliable in and beneficial for Western societies. Nevertheless major breakdowns have occurred, illustrating the complexity of system behavior and of the event sequences which may generate, and showing the negative consequences of dependencies leading to cascading effects.

In the electrical transmission CIs, for example, the analysis of recent major blackouts from 2003 to 2006 (Table 2.2) leads to drawing some conclusions on the main underlying causes and to carving some patterns of common behavior:

- Technical failures (Denmark/Sweden, two independent failures), external impacts (Tokyo, construction work; Brazil, extreme weather conditions) and adverse behavior of protective devices (London) are important triggering events, when not protected by the N-1 security criterion[1] and/or in combination with high-load conditions (Moscow).

[1] Definition of the N-1 security criterion specifies that "any probable single event leading to a loss of a power system element should not endanger the security of the interconnected operation, that is, trigger a cascade of trippings or the loss of a significant amount of consumption. The remaining network elements, which are still in operation, should be able to accommodate the additional load or change of generation, voltage deviation or transient stability regime caused by the initial failure." (Union for the Coordination of Transmission of Electricity 2008).

Table 2.2 Recent major blackouts of electric power supply systems

Blackout		Load loss (GW)	Duration (h)	People affected	Main causes
Aug 14, 2003	Great Lakes, NYC	~60	~16	50 million	Inadequate right-of-way maintenance, EMS failure, poor coordination among neighboring TSOs
Aug 28, 2003	London	0.72	1	500,000	Incorrect line protection device setting
Sept 23, 2003	Denmark/Sweden	6.4	~7	4.2 million	Two independent component failures (not covered by N-1 rule)
Sept 28, 2003	Italy	~30	up to 18	56 million	High load flow CH-I, line flashovers, poor coordination among neighboring TSOs
July 12, 2004	Athens	~9	~3	5 million	Voltage collapse
May 25, 2005	Moscow	2.5	~4	4 million	Transformer fire, high demand leading to overload conditions
June 22, 2005	Switzerland (railway supply)	0.2	~3	200,000 passengers	Non-fulfillment of the N-1 rule, wrong documentation of line protection settings, inadequate alarm processing
Aug 14, 2006	Tokyo	?	–5	0.8 million households	Damage of a main line due to construction work
Nov 4, 2006	Western Europe ("controlled" line cut off)	~14	~2	15 million households	High load flow D-NL, violation of the N-1 rule, poor inter TSO–coordination
Nov 10, 2009	Brazil, Paraguay	~14	~4	60 million	Short circuit on key power line due to bad weather, Itaipu hydro plant (18 GW) shut down

- Organizational factors such as market liberalization and short-term contracting causing operation of the system beyond original design parameters (e.g., Great Lakes, Italy), and stressing operation conditions such as weakening maintenance work and/or inadequate integration of intermittent power generation (e.g., Western Europe) have proven to be outstanding causes.
- As the transmission system operators (TSOs) play a decisive role with regard to contingency management, lack of situational awareness and short-term preparedness, as well as limited real-time monitoring beyond control areas and poorly timed cross-border coordination (e.g., Great Lakes, Italy, Switzerland (rail)) build up as aggravating factors.
- The inadequacy of the N-1 security criterion and, even more importantly, of its inadequate evaluation/implementation in various cases have enforced attempts to make it more stringent and legally binding.

Also, lack of investment due to increasing economic pressure, public resistance, etc., can be observed in many countries and areas, leading to insufficient system monitoring, control, and automation as well as to insufficient grid extension and maintenance (including tree cutting programs [Great Lakes, Switzerland/Italy]), and thus contributing significantly to past blackouts.

As expected, disruption of electricity supply strongly affects our society and other infrastructures which depend on it. The Italian electric power blackout on September 28, 2003 at 3.01 a.m. (Sunday/Monday night) may serve as an example to further elucidate the course of events and delineate the associated consequences:

At the given date, one of the main north–south transit lines through Switzerland—the Lukmanier transmission line—shut down following a flashover between a conductor cable and a tree. This resulted in a redistribution of the electricity in accordance with the laws of physics, and a subsequent overload (110%) of another north–south transit line, namely the San Bernardino transmission line, which due to another flashover also shut down at 3.25 a.m. What followed was a series of cascading failures of other transmission lines in the border region. At that time, the Italian grid was completely separated from the UCTE[2] network. Despite primary frequency control, automatic disconnection of the pump storage plants in Northern Italy, and automatic load shedding (10 GW), the voltage and frequency drop within the Italian grid could not be mastered and generation plants started to trip. This in turn gave rise to a total blackout throughout Italy at 3.27 a.m. (except Sardinia).

After 3 hours, energy was restored in some regions connected to France (such as Liguria). Nine hours later, in the afternoon of September 28, electricity was restored gradually in most places, including Turin, Milan, Venice, and Rome. The energy not supplied due to the rotating outages totaled 12.9 GWh. Rolling blackouts continued to affect about 5% of the population on the next 2 days (September 29–30) as the electricity company ENEL continued its effort to restore supply. Restoring power to the whole country took 18 hours. As a consequence, other infrastructure sectors were affected showing their strong dependence on electricity supply (Fig. 2.1). The effects on the population, the economy and other infrastructures—in greater detail—are given in Table 2.3.

The role of failure cascades and (inter)dependencies among infrastructures (see Chap. 1 and Sect. 2.3) is highlighted in the real examples listed in Table 2.4. Details of the mini telecommunication blackout in Rome, Tor Pagnotta Street, on January 2, 2004 at 5.30 a.m., demonstrate the challenges to (inter)dependency analysis:

Flooding of a Telecom Italia major telecommunication service node occurred when a metallic pipe carrying cooling water for the air conditioning plant broke. The flooding led to several boards/devices failing due to short circuits and the main power supply going out of service. Diesel generators, part of the emergency power supply, failed to start due to the presence of water; only batteries provided power to the boards/devices still working, but finally dropped.

---

[2] Now ENTSO-E.

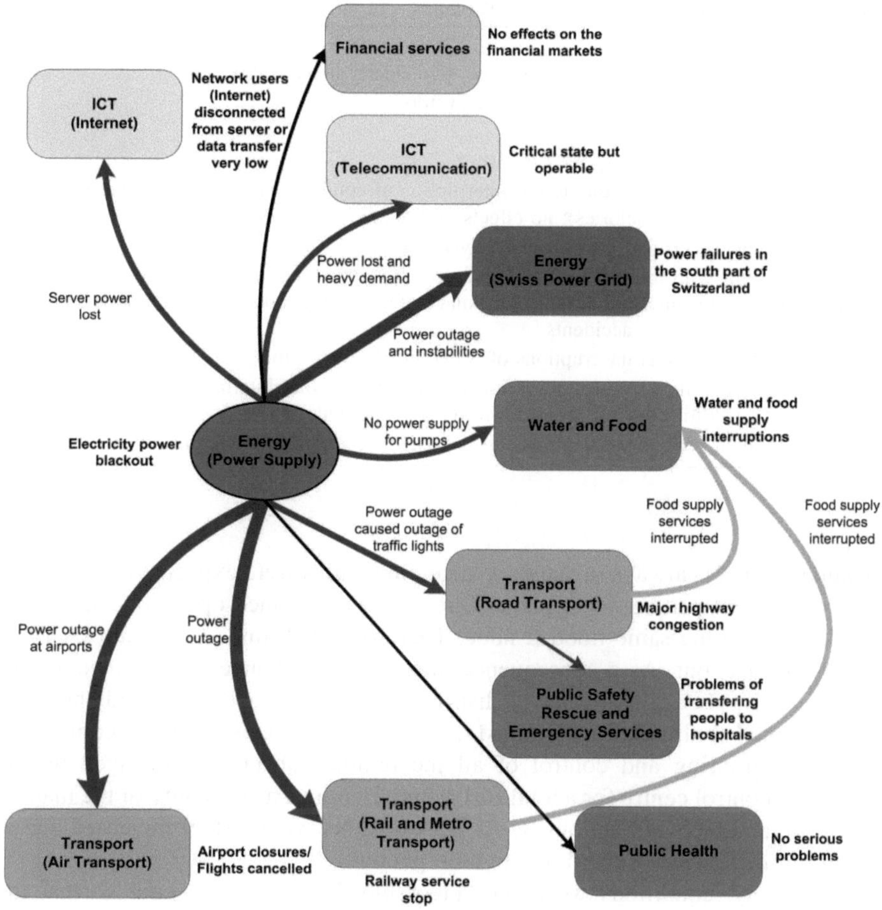

Fig. 2.1 Impact of the Italian blackout on other infrastructure sectors

The Fire Brigade arrived at 7.30 a.m. and worked for pumping out the flooding water and finally individuating the point of the pipe breakage. To start the repair actions, technicians had to shut down the air conditioning plant. The mini blackout caused problems and delays in different infrastructures, including Fiumicino airport, ANSI print agency, post offices and banks, ACEA power distribution and the communication network (both fix-to-fix and fix-to-mobile), connecting the main Italian research institutions (Fig. 2.2).

As mentioned before, the Telco mini blackout also impacted services of ACEA electrical distribution power grid. ACEA has two Control Centres: manned Main Control Centre (Ostiense) and unmanned Disaster Recovery Control Centre (Flaminia). All the tele-measures, commands, and alarms managed by the unmanned control centre are dispatched to the manned one using two redundant TELCO communication links at 2 Mbits/s. One is the main link, the other one a

Table 2.3 Effects of the Italian blackout, September 28, 2003

Impact on the population: strong
About 56 million people have been affected, five elderly persons died
Hundreds of people have been trapped in elevators
Economic losses: moderate
About 120 million € due to spoiled foodstuff
Several hundred thousand € due to the interruption of continuously working industries (e.g., steel, cement or plastic factories); no effects on the financial markets
Impact on dependent critical infrastructures: varying
About 110 trains with more than 30,000 passengers stopped as well as subways in Rome and Milan. Flights were cancelled or delayed. Outage of traffic lights partly led to chaotic situations in major cities, no severe accidents
In some southern regions interruptions of water supply for up to 12 h.
Telephone and mobile networks in a critical state but operable; Internet providers shut down their servers (data transfer rate went down to 5% of normal)
Hospitals without serious problems due to the use of emergency power generators

backup link that is always in stand-by state. Such links were expected to be located on two different geographical paths. Due to a maintenance operation, both links were traversing the same flooded node. Therefore, both links were out of service during the blackout. As a consequence, there was no chance to exchange alarms and signals on the status of power distribution network, and commands between the unmanned centre and the manned one. In such a situation, ACEA completely lost the monitoring and control of all the remote substations managed by the unmanned control centre for a total of 1 h and 23 min. The difficulty of the manual diagnostic and recovery actions by ACEA operators were further increased due to partial out-of-service of fixed and mobile phones.

Fortunately, conditions were very favorable during the blackout so that the power grid required no control actions from ACEA Control Centres to its Remote Terminal Unit, within the duration of the Telco mini blackout.

2.3 Dimensions of Interdependencies

As clearly demonstrated by experienced events, dependencies and in particular interdependencies bear significant practical relevance rather than being a (fairly) new theoretical concept (Rinaldi et al. 2001) introduced six dimensions for their description, and a categorization into four general "types of interdependencies" (Fig. 2.3):

- Physical interdependencies—the state of each is dependent on the material output(s)/flows(s) of the other, e.g., a pipeline network provides gas to fuel a gas-fired power station while the electricity generated is used to power compressors and controls the gas supply network.

Table 2.4 Examples of the importance of interdependencies between critical infrastructures (based on ETH–Laboratory for Safety Analysis 2008)

Event	Duration	Cause	Affected infrastructures	Consequences	Vulnerabilities
May 19, 1998 Failure of Galaxy IV satellite (CRE)	2 days Galaxy IV was not restored	Failure of satellite's primary control processor (technical)	ICT Finance service	45 million customers lost pager service Loss of TV/radio service Disruption of bank card service	Almost all the satellites are not well protected due to their inherent designs and operating environment (design) The replacement of failed satellite is extremely slow and expensive (operational)
July 18, 2001 Baltimore Howard Street Tunnel fire (CIE)	The tunnel entrance was blocked for about 5 days	A freight train detailed while passing through Howard Street Tunnel in Baltimore and caused the fire explosion due to the subsequent ignition of the flammable liquid (technical)	Transport (rail transport) → water and food (geospatial) Transport (rail transport) ↔ ICT (radio) (physical) Transport (rail transport) → energy (power supply) (geospatial) Water and food (water supply) → ICT (Internet/ telecommunication) (geospatial) Water and food (water supply) → energy (power supply) (geospatial)	12 million USD associated with incident 23 bus lines and several train services suspended or delayed Delays of coal and limestone services Extremely heavy road congestion in Baltimore 14 million gallons of water lost (water supply system) 1,200 Baltimore buildings lost electricity Service disruptions for phone/cell phone and	Significance of this rail route was not fully recognized (organizational) Many structures, services, and private utility lines all coexisted within a relatively compact area (design)

(continued)

Table 2.4 (continued)

Event	Duration	Cause	Affected infrastructures	Consequences	Vulnerabilities
				slowed Internet service	
August 23-31, 2005 Hurricane Katrina (CCIE)	Katrina, a category 4 hurricane dissipated 7 days after it formed. Repair and recovery last more than four years	Hurricane Katrina caused severe damage along the Gulf coast from Central Florida to Texas, especially in New Orleans, Louisiana (natural)	Energy (power supply) → Water and food (water supply) (physical) ICT (telecommunication) → public safety, rescues, and emergency service (physical) ICT (telecommunication) → energy (power supply) (logical) Energy (power supply) → ICT (radio) (physical) Transport (road transport) → public health (medical care and hospitals) (physical) Energy (power supply) → public health (medical care and hospitals) (physical) ICT (radio) → public health (medical care and hospitals) (physical)	Damages cost more than 100 billion USD 1,836 fatalities 80% of New Orleans city flooded Drinking water contamination 890,300 customers in Louisiana lost power 30 oil drilling platforms destroyed or damaged 24% of annual oil production and 18% of annual gas production reduced 3 million customers lost phone lines of 2,000 cell sites, out of service Most of major roads in New Orleans area damaged 7 million gallons of oil and 1–2 million gallons of gasoline spilled into southeast Louisiana	The original levees in New Orleans was not prepared to category 4 or 5 hurricane (design) Massive inoperability problems between communication and power supply systems (design/operational) Lack of backup communication devices, such as satellite phone(organizational) Responsibilities (city, state, federal) turned out to be impending (organizational)

(continued)

Table 2.4 (continued)

Event	Duration	Cause	Affected infrastructures	Consequences	Vulnerabilities
				Population in New Orleans reduced half after Katrina	
August 17, 1999 Kocaeli Earthquake (CCIE)	Restoration process lasted from a few days (power supply) to months (transport infrastructure) and up to many years for total recovery	A Mw 7.4 earthquake along the North Anatolian Fault in northwestern Turkey (natural)	IE → ICT (Internet/telecommunication), energy (power supply, oil supply), chemical industry, transport (road/rail), water and food, public safety, rescues, and emergency services, financial services Energy (power supply) → ICT (telecommunication) (physical) Energy (power supply) → transport (rail/road) (physical) Energy (power supply) → water and food (water and food supply) (physical) Energy (oil supply) → transport (road) (physical) Transport (road) → public safety, rescues,	Around 15 billion USD losses Official death toll at 17,127 (unofficial figure closer to 40,000), 43,959 injured, around 500,000 homeless Loss of electricity due to damaged transmission lines and substations Oil sector affected due to damages on industrial plants Various industrial facilities affected including petrochemical plants, pharmaceutical firms, car manufactures, tire companies, paper mills, steel fabrication plants, etc. Loss of water supply services due to	Some heavy populated districts built on ground mainly composed of sea soil which made them vulnerable to any earthquake (organizational/technical) Inadequate seismic design and construction practices lead to collapses (technical) Inadequate earthquake risk management (organizational) Food, clothing, and sheltering efforts generally lagged behind the needs of the people (organizational)

(continued)

Table 2.4 (continued)

Event	Duration	Cause	Affected infrastructures	Consequences	Vulnerabilities
			and emergency services (physical)	failures of pipelines Motorways and railway tracks damaged Telephone service disruptions	
January 2, 2004 Mini Telecommunication Blackout in Rome (CIE)	The communication blackout lasted around 2 h	In the major Telco node, the breakage of a metallic pipe carrying cooling water caused the node failure and cascading failures in the other infrastructures (technical)	ICT (geospatial) Energy (cyber) Transport (cyber) Finance Services (cyber)	ACEA power grid lost the monitoring and control of all the remote substations managed by the unmanned Control Centre Major node failure and all connections connected to the node failed Delays and services perturbations at banks Delays and check-in troubles in Fiumicino airport Delays and services perturbations in ANSI print agency, post offices	The facilities and apparatus in the major Telco node were not checked and maintained carefully and regularly (operational) Because of a maintenance operation, two redundant communication links between two control centers of ACEA power grids were actually routed through the same Telco node. Both links failed too due to geospatial common cause failures (design/ operational) Emergency power

(continued)

Table 2.4 (continued)

Event	Duration	Cause	Affected infrastructures	Consequences	Vulnerabilities
					supply like diesel generators was not kept safely enough and failed to start due to the presence of water (operational)
September 11, 2001 World Trade Center Attack (CCIE)	Part of energy supply and telecommunication were restored within 1 week. More services were restored in 1 year. The complete rebuilding of WTC is still underway	Terrorist attack with two hijacked planes caused the collapse of WTC (malevolent acts)	ICT (geospatial and physical) Energy (geospatial and cyber) Transport (physical and logical) Finance services (geospatial, physical, and cyber) Water and food (geospatial) Public safety, rescue and emergency services (geospatial, physical and cyber) Public health (logical)	Four airplanes crashed *Twin Towers collapsed, destruction of nearby buildings In total, 2,993 people were killed, more than 6,000 injured Loss of power supply in a big area Loss of gas and steam supply A lot of phone lines and data lines damaged, most communication traffic rerouted New York Stock Exchange closed,	Geospatial concentrations of critical infrastructure may be distinctly vulnerable to catastrophic disruption (design and operational) Lack of interoperability results in police and fire not being able to communicate with one another during an emergency (organizational and operational) Revealed shortfalls in the emergency

(continued)

Table 2.4 (continued)

Event	Duration	Cause	Affected infrastructures	Consequences	Vulnerabilities
				international economy affected Loss of water supply All air and subway services suspended	services sector being able to respond to large-scale terrorist incidents and other catastrophes disasters that required extensive cooperation among local, state, and federal emergency response organizations (organizational and operational) Communications and business continuity will further deteriorate after the initial impact of a disaster if backup power generation facilities are not provided with guaranteed access to fuel and maintenance (organizational)

CRE cascade resulting event, *CIE* cascade initiating event, *CCIE* common cause initiating

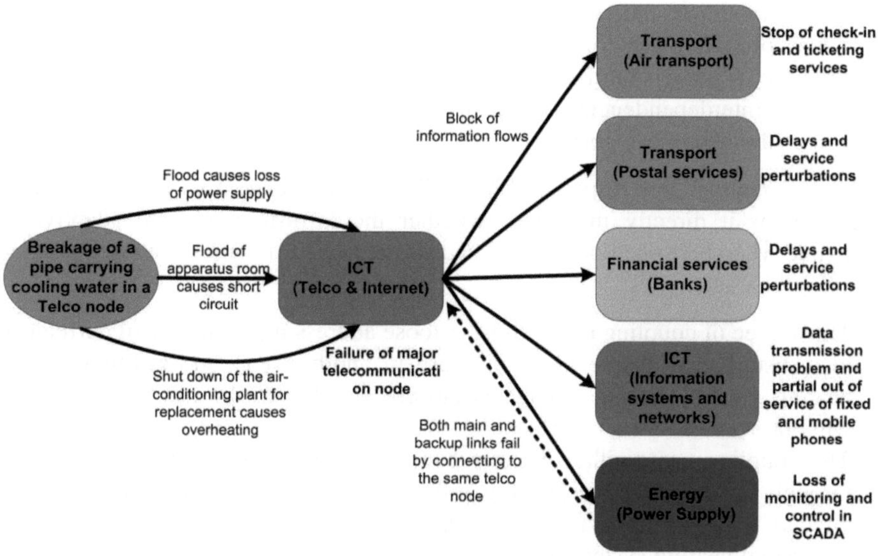

Fig. 2.2 Infrastructures affected by the mini telecommunication blackout in Rome, 2004

Fig. 2.3 Dimensions for describing infrastructure interdependencies (Rinaldi et al. 2001)

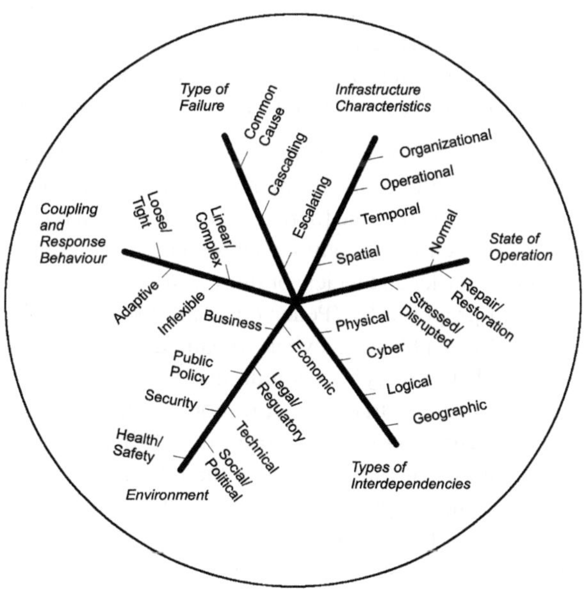

- Geographic interdependencies—elements are in close spatial proximity and a local environmental event affects components across multiple infrastructures, e.g., earthquake, flooding or a fire.
- Cyber interdependencies—connect infrastructures to one another via electronic, informational links, e.g., a supervisory control and data acquisition (SCADA)

system monitors and controls elements of the electric power grid—likewise, it may provide pieces of information or intelligence supporting another infrastructure or a decision-making process elsewhere.
- Logical interdependencies—exist between infrastructures that do not fall into one of the above categories.

The "coupling[3] and response behavior" of interdependent CI deserves special attention, as it directly influences whether the infrastructures are adaptive or inflexible when perturbed or stressed. Rinaldi et al. (2001) introduce three primary coupling characteristics:

- The degree of coupling either tight or loose addressing the nature of correlation of a disturbance in one agent to those in another; e.g., the gas-fired spatial heating system without storage is closely coupled to the gas supply system without "time to give" or slack.
- The coupling order either directly connected (first-order effect) or indirectly through one or more intervening infrastructures (second-order up to *n*-order effects); e.g., loss of electric power may directly affect the pumps and control of the spatial heating system directly and indirectly affects the fuel supply via the compressors of the gas supply system, if they are electrically driven.
- The linearity or non-linearity/complexity of the interaction; i.e., whether or not agents can interact with other agents outside the normal scheme, or production, or operational sequence, not intended by design as being subtle and difficult to detect, showing unfamiliar feedback loops; e.g., a large scale areal event such as extreme heat affecting various agents simultaneously.

Figure 2.4 depicts all this by taking the prolonged electric power problems in California as the basis. Elements of other dimensions, in particular "environment" (business and economic [deregulation], legal and regulatory [public policy]) and "type of failure" (subset of common causes) might be added to the four above categories of types of interdependencies, or used to specify "logical interdependencies". The latter has been proposed by (Pederson et al. 2006) who slightly expanded the above taxonomy from physical, cyber (renamed informational), and geographic (renamed geospatial) to policy/procedural, and societal interdependencies. These types of interdependencies carry a state, or consequences from events in one infrastructure to other infrastructures, although no direct linkage or relationship in a physical sense exists; halt of air traffic for more than 24 h in the US and air traffic drop worldwide following the "September 11 attack" may serve as reference examples.

Relating to physical-engineered CIs, the six dimensions proposed by (Rinaldi et al. 2001) still seem to be appropriate to facilitate the identification, understanding, and analysis of interdependencies as well as of dependencies, and to frame the requirements for modeling and simulation approaches. A multiple/combined (rather than "silver bullet" single) approach is needed to address, in a

[3] According to *Webster Dictionary*: The act of bringing or coming together. (Mech.) A device or contrivance which serves…. to connect adjacent parts.

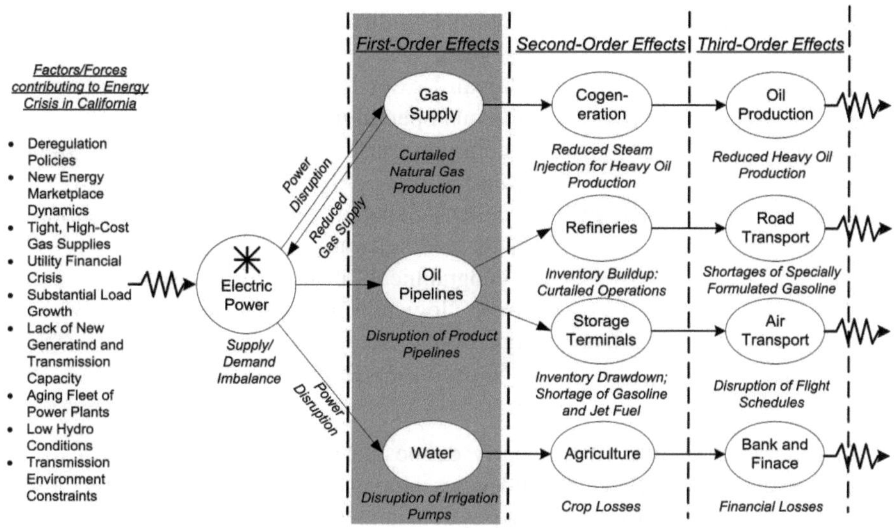

Fig. 2.4 Examples of nth-order interdependencies and effects (Rinaldi et al. 2001)

Fig. 2.5 Dimensions for describing infrastructure interdependencies (according to (Rinaldi et al. 2001), modified by authors in italic)

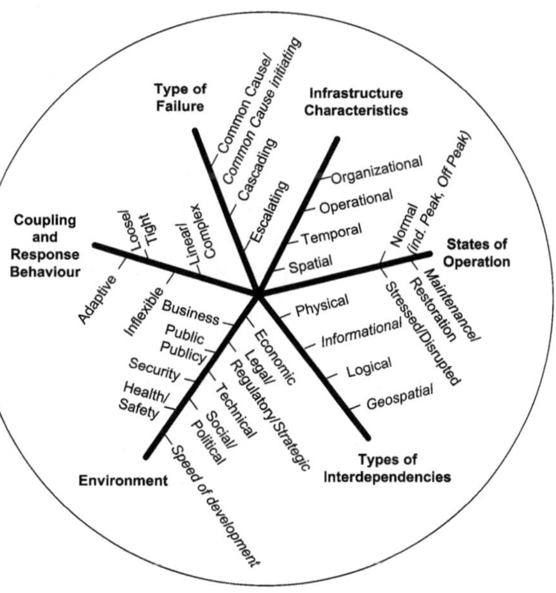

consistent manner, all of these interrelated factors and system environment classes/ attributes, respectively.

The following extensions to the six dimensions and related elements are here proposed (see Fig. 2.5 as a modification of Fig. 2.3) to strengthen their representation:

"*State of operation*":	"normal"—distinction between nominal, peak, off peak "repair" extended to maintenance during continuous operation or down states
"*Type of failure*":	"common cause initiating" added to "common cause"
"*Types of interdependencies*":	"cyber" changed to informational including hard- and software "geographical" changed to geospatial Note "logical" includes lacking diversity, functional, etc.
"*Environment*":	"Speed of developments/changes" added

In general, it is difficult to clearly define the boundaries of individual infrastructures and model (inter)dependencies among them adequately. Often, infrastructures are decomposed into a physical and supporting/controlling part and modeled in a linear fashion. Let us take the electrical power grid (physical system under control) and the monitoring and control (SCADA) system as examples. Only if the SCADA system is dedicated, does not make use of commercial systems to transfer data and commands, e.g., the open Internet, and does not incorporate common hardware and software, then it can be modeled as part of the electric power infrastructure, including dependencies within. If not, as it is obviously the case in many countries, it would be closely linked to the information and communication infrastructure and must be modeled in a less simplistic way, "not to overlook the true complex nature of interconnected infrastructures" (Rinaldi et al. 2001). Taking this into account, many vulnerability analysts call for a "system-of-systems" approach.

Failures (negative impact) that arise from (inter-)dependencies can be classified as follows:

(1) One event causing failure or loss of service of more than one infrastructure, such as areal external events (earthquakes, floods, extreme weather conditions, etc.), due to spatial proximity (called *common cause initiating events*).
(2) Failure of one infrastructure causing failure or loss of service of at least another infrastructure, e.g., rupture of mains of the water supply system (called *cascade initiating events*).
(3) Failure or loss of service resulting from an event in another infrastructure, e.g., failure of gas lines due to loss of main electricity supply if compressors are electronically driven (called *cascade resulting events*).
(4) Failure or loss of service of one infrastructure escalating "domino effect" because of failure of another affected infrastructure, e.g., failure of the electric power system leading to failure of the SCADA system and by this affecting restoration of the electric power system (called *escalating events*).

Events being neither one of these four types maybe called independent. The types of non-independent events are not mutually exclusive.

2.4 Empirical Investigations on Critical Infrastructure (Inter)Dependencies

As outlined before, interdependencies within and among CIs are recognized as both opportunities, e.g., increased coping capacity and elements of increased vulnerability. With regard to the latter, empirical studies have been made and published focusing on building databases of different kinds and/or to obtain findings for decision making.

For example, the database in (Luiijf et al. 2009) was built from public reports of disruptions of CIs from open sources like newspapers and Internet news outlets. The following results were derived from a subset of 1,749 failure incidents in 29 European countries (95% occurred after 2000) with noticeable effect to society (e.g., at least 100,000 electric power customers affected). Events have been classified as "cascade initiating", "cascade resulting," and "independent" (see previous section). The results disclose that:

- "Cascades resulting" events are more frequent than anecdotally thought, i.e., almost 30% (501 out of 1,749) of the reported incidents result from incidents in other services, and that "cascade initiators" are about half as frequent (268 out of 1,749, see Table 2.5).
- The dependency matrix is sparsely populated and cascades are highly asymmetrical; the dependencies are more focused and directional than often thought.
- Energy and telecommunication are the main cascading initiating sectors, energy is the only sector which initiates more cascades than it ends up receiving (146 versus 76, see Table 2.5).

Table 2.5 Categorization of number of CI disruption events (Luiijf et al. 2009)

CI sector	Cascade initiating	Cascade resulting	Interdependent	Total	Sample size
Education	0	3	1	4	4
Energy	146	76	388	609	590
Financial Services	1	26	33	60	60
Food	0	4	3	8	8
Government	2	40	26	68	67
Health	1	16	22	39	39
Industry	5	15	7	27	27
Internet	15	51	95	161	160
Postal Services	1	0	0	1	1
Telecom	69	125	114	308	295
Transport	19	128	276	423	433
Water	9	18	51	78	76
Total	268	501	1,017	1,786	1,749

Table 2.6 Events categorized by initiating and affected sector (number of events) (Luiijf et al. 2009)

	Initiating sector											
CI Sector	No sector	Energy	Financial services	Government	Health	Industry	Internet	Postal services	Telecom	Transport	Water	Grand total
Education	1	1									2	4
Energy	515	65				4			2	1	3	589
Financial Services	34	5	3				3		15			60
Food	4	3								1		
Government	27	17		1	1	1	4		14	1	1	67
Health	23	11			2				2		1	39
Industry	12	12				1				1	1	27
Internet	109	14					10		27			160
Postal Services	1											1
Telecom	170	62					1		57	5		295
Transport	294	98		1		3		1	5	15	5	422
Water	58	14				2					2	76
Total	1,248	302	3	2	3	11	18	1	122	24	15	1,749

Table 2.7 Elements for criticality definition

Conditions for a critical situation	Criticality criteria
"If disrupted or destroyed"	Serious "debilitation and impact on", "would incapacitate", "so vital to", "essential to":
"The incapacity of destruction of such systems and assets"	The "entire system"
"If degraded or rendered unavailable for an extended period"	The "national public health", "safety"
"Disabling any of them"	"National security"
(Non)"continuous, reliable operation"	"National defense"
"Disturbance, functional deficiency or destruction"	"National economic security"
	"Minimum operations of the economy and the effective functioning government"
	"Social or economic well-being of citizens or of the nation"
	"Quality of life"
	"Or any combination of those matters"

- Within the energy sector 61 (out of 65) dependencies exist between the electrical power subsector services, and within telecommunication services disruptions of telecom backbones most seriously affect Internet services (see Table 2.6).

With regard to escalation of cascades, the analysis shows that a cascade initiating event in the energy sector triggers 2.06 disruptions of other services, but taking all events into account (including independent) only one out of two events triggers a disruption of another CI. For the telecommunication sector the respective numbers are 1.86 disruptions and two out of five. Interestingly, 421 events (out of 501 resulting in cascades) are first level, 76 are second level, and 4 are third level cascades; no deeper cascades have been found. This contradicts the results of the evaluation of selected experienced events (Table 2.4) and may be due to a bias of the open source media information consulted.

Another fairly simple measure of (inter)dependency was proposed by Zimmermann (2004) which compares the duration of outages in the initial system disruption, e.g., power outage, with the duration of outages of specific public services and businesses affected. The investigations based on power outage data from North America from 1990 through 2004. Results showed that the duration of outages linked to the electricity outage for affected public services exceeded the duration of the initial power outage itself pointing to the fact that cascading events did escalate.

2.5 Degree of Criticality

The definition(s) of CI (this chapter) focus on systems and assets which are considered critical with regards to some criteria, possibly varying from one country/region specific definition and associated perspective to another. These

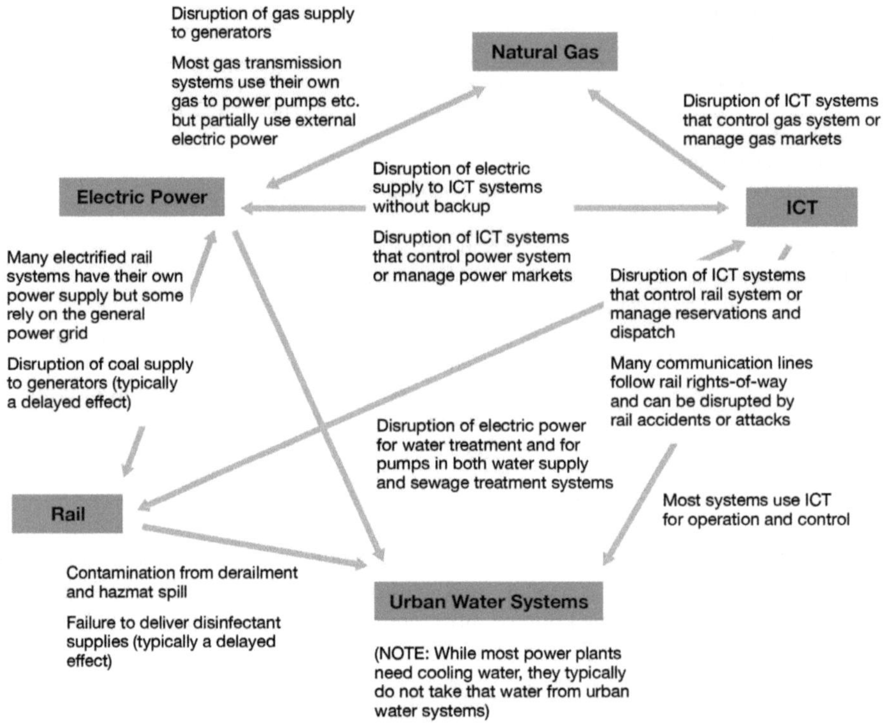

Fig. 2.6 Example for the evaluation of the degree of criticality (IRGC 2006)

definitions (Table 2.7) are descriptive rather than a precise yardstick to objectively assess criticality and its degree. There is no standardized usage and broad-based mutual understanding of what criticality is and how to measure it (see also Bouchon 2006). Nevertheless, the description of conditions for a critical situation focuses on a disturbance or loss of continuous (reliable) service and places the analysis of CIs in the realm of the analysis of reliability (systems view) or availability (users view), of risks for the owner/operator and/or the public due to adverse events, and of vulnerability of the system and/or the society.

Infrastructures are considered highly critical because of being both a trigger of a potential crisis and a means to resolve a crisis/emergency situation; the telecommunication systems may illustrate this. Therefore, criticality objectives are primarily related to strategic objectives of a state entity although other subjective standpoints, i.e., of owners/operators, insurers, other stakeholders and the general public, are worth mentioning and may lead to other criticality criteria. Referring to the Swiss Federal program for critical infrastructure protection (CIP) as an example, the "criticality of an infrastructure" denotes its relative importance regarding the potential consequences of a disturbance, functional deficiency, or destruction for the public and its vital resources" (Federal Office of Civil Protection 2009). The probability of such an event is deliberately not regarded as important.

Attempts have been made to further specify criticality and to distinguish degrees of criticality. The European Commission—EC (EC 2004) has proposed to distinguish three criteria—within its concept of CIP focused on the fight against terrorism:

- *Scope*: The loss of a critical infrastructure element is rated by the extent of the geographic area which could be affected by its loss or unavailability, i.e., international, national, provincial/territorial, or local.
- *Magnitude*: The degree of the impact of loss can be assessed as none, minimal, moderate, or major. Among the criteria which could be used to assess potential magnitude are:

  (a) Public impact (amount of population affected, loss of life, medical illness, serious injury, evacuation)
  (b) Economic (GDP effect, significance of economic loss and/or degradation of products or services)
  (c) Environmental (impact on the public and surrounding location)
  (d) Interdependency (among other critical infrastructure elements)
  (e) Political (confidence in the ability of government)

- *Effects of time*: This criteria ascertains at what point in time the loss of an element could have a serious impact (i.e., immediate, 24–28 h, 1 week, other).

This has been taken up by the International Risk Governance Council (IRGC) to assess the degree of criticality of five coupled physical-engineered infrastructures. This was done semi-quantitatively, based on expert judgment and screening analysis applying the so called traffic light model (IRGC 2006); see Fig. 2.6 for a snapshot of the results. Sometimes it might be of interest to distinguish between objectives, e.g., economy/economic security, public health and safety, and to address interdependencies separately.

Taking again the Swiss CIP program as the example, it was proposed to assess the criticality at the level of 31 subsectors[4] 'according to their "relative importance" (see definition above). The main purpose is to give steer to more detailed analyses aiming at the identification of critical elements in prioritized infrastructures (called vertical criticality) and to strategic planning (BABS 2008). Three criteria are distinguished:

- Effect on other subsectors (dependence)
- Effect on the public
- Effect on the economy

Four categories have been established to specify the effect (consequences)—from none (0), small (1), median (2) to large (3). It is assumed that the "loss of continuous service" will take place without pre-warning, will last approximately 3 weeks, and will affect the whole country. A spider diagram is used to illustrate

[4] The sector energy, for instance, is divided into the subsectors electric power, oil, and gas supply.

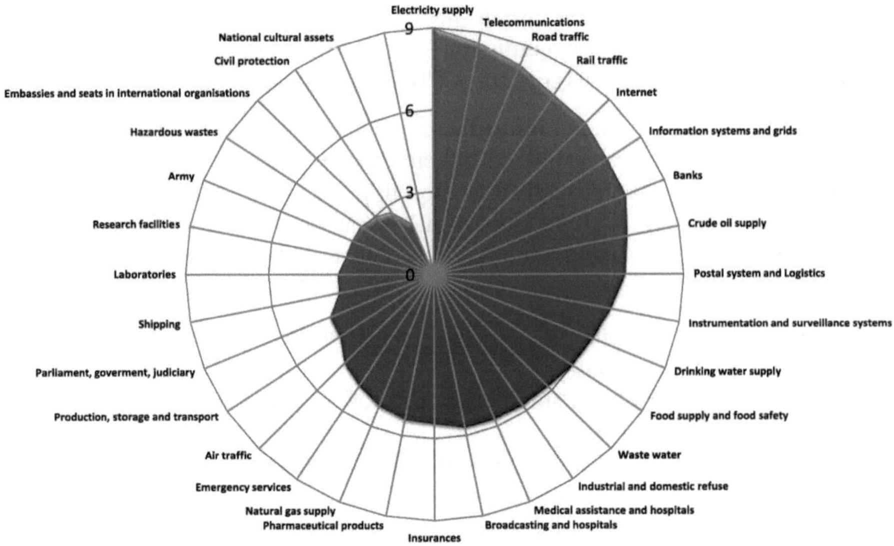

Fig. 2.7 Criticality of subsectors (Federal Office of Civil Protection 2008)

the results (Fig. 2.7): electric power supply and telecommunication are highest ranked.

Activities are undergoing at the level of the European Union aimed at identifying "European critical infrastructures" (EC 2008) by taking three aspects into account:

- Risk of casualties (number of fatalities and/or injuries)
- Economic loss (percentage of GDP)
- Public effect (number of people affected)

To be regarded "critical" for the European Union at least two Member States must be significantly affected by a loss of service of the infrastructures under consideration (currently the energy [electric power, oil and gas supply] and transport sectors). As in all other cases known to the authors, definition and assessment of criticality focus on loss of service; misuse of infrastructure to intentionally cause harm to the public, economy, and government ("weaponizing") is not taken into consideration; penetration into unmonitored parts of the urban drinking water system and dumping of hazardous substances may serve as fictitious example for illustration.

References

Bouchon S (2006) The vulnerability of interdependent critical infrastructures systems: epistemological and conceptual state-of-the-art (EU report). EU Commission, Joint Research Centre, Ispra, Italy

Dueñas-Osorio L, Vemuru SM (2009) Cascading failures in complex infrastructure systems. Struct Saf 31:157–167

Dueñas-Osorio L, Vemuru SM (2009) Cascading failures in complex infrastructure systems. Struct Saf 31:157–167

ETH–Laboratory for Safety Analysis (2008) Interdependencies. Report for FOCP, 2009

European Commission (2004) Critical infrastructure protection in the fight against terrorism, COM (2004) 702 final, Brussels, 20 October 2004

European Commission (2008) Directive on the identification and designation of European critical infrastructure and the assessment of the need to improve their protection. Council of the European Union, Brussels

Federal Office of Civil Protection (2008) Schlussbericht horizontale Kritikalität (final report on horizontal criticality), Berne, 10/2008 (for internal use).

Federal Office of Civil Protection (2009) The Federal Council's basic strategy for critical infrastructure protection, Berne, 5/2009

Guckenheimer J, Ottino JM (2008) Foundations for complex systems research in the physical sciences and engineering. Report from an NSF Workshop in September 2008

IRGC (2006) Managing and reducing social vulnerabilities from coupled critical infrastructures, White Paper No. 3, International Risk Governance Council, Geneva, p 68

Luiijf E, Nieuwenhuijs A, Klaver M, van Eeten M, Cruz E (2009) Empirical findings on critical infrastructure dependencies in Europe. In: Proceedings of CRITIS 2008, Rome, Italy, 13–15 October 2008, pp 302–310

Pederson P, Dudenhoeffer D, Hartley S, Permann M (2006) Critical infrastructure interdependency modeling: a survey of US and international research. Technical report INL/EXT-06-11464. Idaho National Laboratory, Idaho, USA

Rinaldi SM, Peerenboom JP, Kelly TK (2001) Identifying, understanding, and analyzing critical infrastructure interdependencies. IEEE Contr Syst Mag 21(6):11–25

Union for the Coordination of Transmission of Electricity (2008) Operational handbook. www.ucte.org. Accessed date/month/2008

Zimmermann R (2004) Decision-making and the vulnerability of interdependent critical infrastructure. In: Proceedings of the IEEE international conference on systems, man, and cybernetics, the Hague, Netherlands

Chapter 3
Challenges to Methods for the Vulnerability Analysis of Critical Infrastructures

As has been discussed in the previous section (see Table 2.1) critical infrastructures (CIs) exhibit a number of complex system characteristics which call for analyzing them as a whole and make their holistic study highly challenging. A comprehensive vulnerability analysis requires not only the consideration of a large number of spatially distributed, interacting elements with nonlinear behavior and feedback loops, but also a broad spectrum of hazards and threats including failures and threats. It comprises three main activities (Fig. 3.1):

- System analysis including system properties (e.g., physical and logical structures and operation modes)
- Quantification of system vulnerability indicators and identification of important elements
- Application to system improvements either technical or organizational

The challenges to methods for vulnerability analysis depend on the objectives of the assessment and on the system characteristics. More specifically, the following main challenges are to be tackled, while a look at how different methods basically respond to them is given in Chap. 4.

3.1 Emergent Behavior

The local interaction of a plethora of system components often results in global behavior, which is difficult or even counterintuitive to anticipate. In general, such hard-to-predict collective phenomena are referred to as "emergent behavior" (e.g., Bonabeau 2002). As a consequence, vulnerability analysis methods have to be developed and applied in such a way that emergent phenomena, potentially crucial for the overall robustness of critical infrastructures, are sufficiently captured.

According to Wolf and Holvoet (2005), the emergence properties include micro–macro effect, radical novelty, coherence, dynamics, interacting parts,

W. Kröger and E. Zio, *Vulnerable Systems*, DOI: 10.1007/978-0-85729-655-9_3,

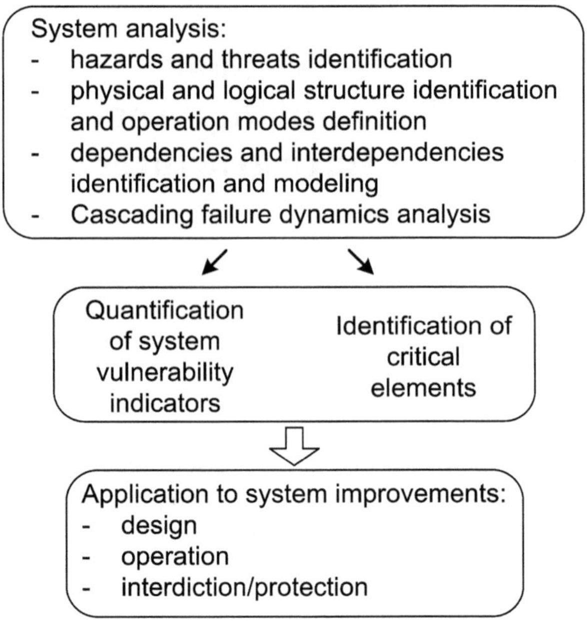

Fig. 3.1 Conceptualization of CI vulnerability assessment

decentralized control, two-way link and flexibility. For an individual infrastructure, the definition of the micro-level (lowest level of the modeling) depends on the level of detail, which has to be chosen carefully. Furthermore, a careful decision needs to be made with regard to the trade-off between two extreme viewpoints: so called non-reductionism, which means that the macro-level emergence is not reducible to the micro-level parts of the system, and the opposite opinion that emergent properties can be studied by taking a system apart and looking at the parts (so-called reductionism) and that the behavior of the system can be reduced to that of its elements (Kubik 2002).

3.2 Intricate Rules of Interaction

The required performance of critical infrastructures relies on intricate, often nonlinear interactions among a large number of interconnected and geographically distributed components of different types, including both technical and non-technical elements. Even their interaction with the regulatory, legal or institutional framework may eventually affect the overall vulnerability of infrastructure systems. Furthermore, the normal operation of these systems does not allow to detect ‘hidden’ interactions which might become crucial for the evolution of (cascading) failure events; the dynamic degradation of networks is sensitive to parameter variations and the system behavior cannot be described by linear equations.

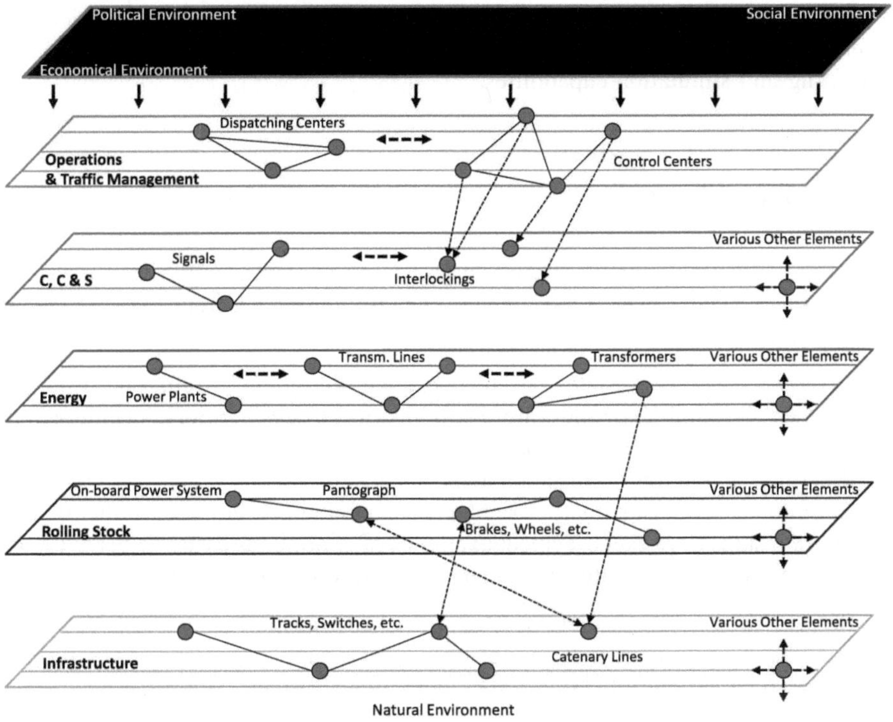

Fig. 3.2 Illustration of railway system as a multi-layer infrastructure network

Most CI are open systems, i.e., they exchange with the environment very intensively. Each CI needs input in terms of energy, resources, information, etc.; the services provided are the system outputs. The interactions between the system and its environment have to be taken into account carefully and the undergoing exchanges to be established adequately via interfaces.

3.3 Single System Features and "Systems-of-Systems"

3.3.1 Multi-Layered Systems

Single critical infrastructures can be viewed as large-scale systems built up from various co-existing techno-socio-economic layers. For instance, transportation systems or electric power supply infrastructures consist of physical networks, communication and control layers, among others, all embedded within a market-driven organizational framework. Figure 3.2 shows one way of illustrating the different layers making up the overall railway system (see also Fig. 1.2 for further explanations).

Tackling the system fragility induced by these layered (inter)dependencies, even when only within one CI, remains one of the main challenges of today's modeling and simulation capabilities.

3.3.2 State Changes

Modeling and simulation of dynamical processes emulate changes of the system/components states over time which can be either discrete or continuous (Borshchev et al., 2002):

1. *Discrete processes:* State variables change "jumping" from one value to another. Typical examples are break-down failures, factory operations or shipping facilities, in which the material or information being simulated can be described as moving in discrete steps or packets, etc.
2. *Continuous processes:* State variables are changing continuously, rather than in discrete steps or packets. Typical examples of continuous dynamic behavior are wear-out processes, water movement through reservoirs and pipes, etc.

Besides the typical discrete systems (clocked transport systems, telecommunication) and continuous systems (electric power supply, gas and water supply) there are so called hybrid systems, e.g., *continuous systems controlled by discrete events,* which include time-dependent changes and jumping events. An example is the continuous electricity supply provided by the electric power system affected by a shutdown event initiated by the SCADA-system in the case of line overload.

In the case of discrete state changes, models must treat entities (e.g., units of traffic), resources (elements that service entities), and control elements (elements that determine the states of the entities and resources). In the case of continuous processes, the model has to include differential or integral equations that describe the evolution of the system. For hybrid systems, a separation of the underlying discrete and continuous processes into two different layers is typically required.

When modeling these changes we have also to be aware of multiple time scales. For example, within the electric power transmission system switching-over voltages happen within microseconds whereas, e.g., some maintenance actions may be scheduled once a year and take hours to days.

3.3.3 Evolving Systems

Modern societies have become embedded in an increasingly dense web of interconnected infrastructures which are subject to a continuous and fairly rapid evolution over time characterized by, for instance, geographic expansions, integration of new technologies, coexistence of old and new technologies or changing market conditions. This evolution at different time and spatial scales brings about

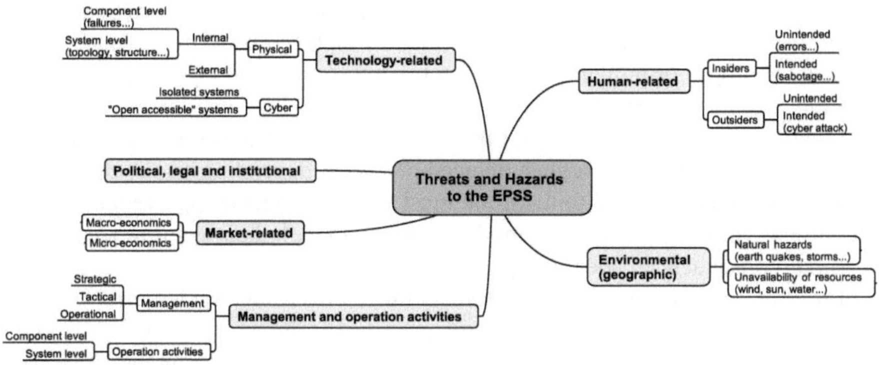

Fig. 3.3 Spectrum of multiple hazards and threats to CI (Kröger 2008)

the need to continuously adapt and update vulnerability assessments. Such a process, however, might become a prohibitive task.

3.3.4 "System-of-Systems"

Modern infrastructure networks are strongly coupled to each other by different types of (inter)dependencies (see Sect. 2.3) of different degrees. Therefore, they should preferably be modeled as interconnected "system-of-systems".[1] A fundamental property of interdependent networks is that failures within one network may cascade through dependent nodes in other networks (Buldyrev 2010). Powerful methods are needed to describe the behavior of such a system as a whole (not as a sum of single systems).

3.4 Broad Spectrum of Hazards and Threats

In principle, the model of interconnected or single CI must consider all relevant hazards and threats. The risks or vulnerabilities related to critical infrastructures often refer to being *systemic* by nature. "Investigating systemic risks or vulnerabilities goes beyond the usual agent-consequence analysis and focuses on interdependencies and spill-over risk clusters" (IRGC 2005). Hence, approaches are required to capture this holistic view.

[1] No universally accepted definition of the term exists; we focus on the following one (DeLaurentis 2007): "A system-of-systems consists of multiple, heterogeneous, operationally distributed, occasionally independently operating systems embedded in networks at multiple levels that evolve over time.

Depending on the goal of the analysis a decision has to be taken about the spectrum of hazards and threats to be included which can be of many-sided nature, either technology- or human-related, natural and operational as well as contextual (see Fig. 3.3 for illustration).

Technology-related hazards and threats include physical failures at component and system level as well as cyber-induced problems depending on the openness of the system. Challenges are to distil the most important technology-related hazards and threats and related failure modes, and to model their occurrence and the potential consequences adequately.

The spectrum of *human-related* threats ranges from unintended errors, e.g., of operators (a) of not-performing the required task ("errors of omission") or (b) doing something else instead ("errors of commission") to (c) intended "errors" (sabotage) made by insiders or outsiders to (d) targeted malicious attacks, either physical (e.g., explosive devices) or cyber. While challenges resulting from (a) are well known and call for models to deal with human reliability, (b) to (d) call for even more sophisticated models capable of describing a complex human behavior under abnormal conditions, together with systems response models to predict a potential damage to the system analyzed.

The wide-area, large-scale nature of critical infrastructures makes them highly susceptible to *natural* hazards, such as earthquakes, flooding, extreme weather conditions including heat, cold, storms, etc. Special models are necessary to quantify hazards (e.g., seismic hazard curves) and to capture the impact on exposed systems. Other environmental hazards relate to, e.g., geographic proximity of various components and systems, challenging models to deal with interdependencies and cascading effects.

Inadequate *management and operational* activities can threaten infrastructures, in particular errors in instructions and norms can lead to service interruptions or even cause the complete failure of a system; it is a non-trivial task to find out such weaknesses in the documentation and integrate them into the dynamic model of a system.

Further hazards and threats relate to factors which are often called *contextual* such as market-related as well as political, legal or institutional. There is evidence that liberalization of the market and associated political decisions have cut off reliability factors such as reserves and redundancies and slowed down investments while integration of the systems has been further increased (IRGC 2005). Therefore, contextual factors clearly affect system vulnerability in complex ways. They add challenges to system modeling and analysis (simulation). There is a clear need to take increased economic pressure into account when assessing human response behavior, to modify maintenance strategies and spare-parts managing, to focus on the search for entry points for hacker attacks after the industrial control systems have become more open and less dedicated, etc.

The consideration of a broad spectrum of hazards and threats, often called "all hazards approach", including failures and failure combinations of highly reliable elements, and the occurrence of rare natural hazardous events may cause specific modeling difficulties. The simulation of such events, of low probability of occurrence, brings issues related to the time frame of the analysis and to the number of simulation runs needed to observe the effects of a statistically significant number of rare events (very low by nature).

References

Bonabeau E (2002) Predicting the unpredictable. Harvard Bus Rev 80(3):5–11

Borshchev A, Karpov Y, Kharitonov V (2002) Distributed simulation of hybrid systems with AnyLogic and HLA. Future Gener Comp Sy 18(6):829–839

Buldyrev SV, Parshani R, Paul G, Stanley HE, Havlin S (2010) Catastrophic cascade of failures in interdependent networks. Nature 464:1025–1028

DeLaurentis D (2007) Role of humans in complexity of a system-of-systems. In: Duffy VG (ed) Digital human modeling, LNCS 4561. Springer-Verlag, Berlin, Heidelberg, pp 363–371

IRGC (2005) Risk governance: towards an integrative approach. White Paper No. 1, IRGC, Geneva

Kröger W (2008) Critical infrastructures at risk: a need for a new conceptual approach and extended analytical tools. Reliab Eng Syst Safe 93:1781–1787

Kubik A (2002) Towards a formalization of emergence. Artif Life 9(1):41–65

Wolf TD, Holvoet T (2005) Emergence versus self-organization: different concepts but promising when combined. In: Proceedings of the second international workshop on engineering self-organizing applications (July 2004), Springer-Verlag, Berlin, Heidelberg, pp. 96–110

Chapter 4
Basic Approaches

The two main outputs of a vulnerability assessment of critical infrastructures (CIs) are the quantification of system vulnerability indicators and the identification of critical elements (Chap. 2). The information they provide is complementary: while vulnerability indicators are parameters encompassing the static and/or dynamic characteristics of the whole system, the identification of critical elements comes from their ranking with respect to their individual connectivity efficiency and/or their contributions to the propagation of failures, with their effects, through the network.

A number of approaches can be undertaken for the vulnerability assessment of CIs depending on the type of system, the objective of the analysis and the available information. In this chapter, an attempt is made to defining the characteristics of these approaches, framed into categories of homogeneity (Table 4.1, including references), and to explaining them briefly. Those methods regarded as most important or promising by the authors and little described in the literature are explained in detail in Chap. 6.

4.1 Statistical Analysis

The extensive and growing use of information and telecommunication (IT) systems to capture data about CIs operation and performance (e.g., traffic delays in a transportation system, loss of power in an electric power network and signs of electronic intrusion in a banking system) provides rich data sets which can support vulnerability analyses. However, using these data effectively is difficult for a number of reasons: (i) data about CIs operation and performance generally come from a variety of past operating conditions that may not fully reflect the situations of interest at present and in the future; (ii) the relationships between the measures of the operating conditions (e.g., the loads placed on the different portions of the system) and system performance may be complicated and poorly understood; (iii) the data sets may be very large, making it difficult to draw clear insights

W. Kröger and E. Zio, *Vulnerable Systems*, DOI: 10.1007/978-0-85729-655-9_4,

Table 4.1 Categorization of the approaches for the vulnerability assessment of CIs

Approaches		Assessment steps			Outputs		Practical objectives		
	Techniques	Logic structure identification	(Inter)dependencies identification and modeling	Cascading failure analysis	System vulnerability indicators	Critical elements	Design optimization	Operation optimization	Interdiction/ protection optimization
Statistical analysis	Statistical models	No: aggregate data considered (Dekker 2005; Yamijala et al. 2009; Debon et al. 2010); different statistical models for different scenarios (Lord et al. 2005)	No	No	Yes (Dekker 2005; Yamijala et al. 2009; Debon et al. 2010)	Yes: identification of the critical parameters (Debon et al. 2010)	Yes (Debon et al. 2010)	No	No
Probabilistic modeling	Markov chains	Yes: but require enumeration of all possible combinations of components' states (Iyer et al. 2009)	No: single infrastructures are considered as monolithic entities (Sultana and Chen 2009)	Yes (Iyer et al. 2009)	Yes (Iyer et al. 2009; Guida et al. 2010)	Yes: severity of distur-bances can be obtained (Augutis et al. 2010)	No	No	No
	Markov/Petri nets	Yes: when combined with network structure (Langeron et al. 2010)	Yes: but only high level qualitative analysis (Krings and Oman 2002; Laprie et al. 2007; Sultana and Chen 2009)	No: high level qualita-tive analysis (Laprie et al. 2007)	Yes (Langeron et al. 2010)	No	No	No	No
	Probabilistic dynamics modeling	No: analytical model (Watts 2002; Dobson et al. 2005)	Yes (Newman et al. 2005)	Yes (Coffman et al. 2002; Buzna et al. 2007)	Yes (Dobson et al. 2004)	Yes: identification of the critical parameters (Coffman et al. 2002; Chen et al. 2005)	No	No	Yes (Chen et al. 2005; Buzna et al. 2007; Anghel et al. 2007)
	Bayesian networks	Yes (Doguc and Ramirez-Marquez 2009)	No	No	Yes (Doguc and Ramirez-Marquez 2009)	No	No	No	No

(continued)

Table 4.1 (continued)

Approaches		Assessment steps			Outputs		Practical objectives		
	Techniques	Logic structure identification	(Inter)dependencies identification and modeling	Cascading failure analysis	System vulnerability indicators	Critical elements	Design optimization	Operation optimization	Interdiction/ protection optimization
Risk analysis, hazop, FMEA	Quantitative assessment (PSA)	Yes (Apostolakis and Lemon 2005; Koonce et al. 2008)	Yes (Apostolakis and Lemon 2005)	Yes: using load flow simulation (Koonce et al. 2008)	Yes (Paté-Cornell and Guikema 2002; Apostolakis and Lemon 2005)	Yes (Koonce et al. 2008)	No	No	Yes (Paté-Cornell and Guikema 2002; Flammini et al. 2009)
	Tabular methods/ expert judgment	Yes (Moore 2006; Piwowar et al. 2009)	Partially: qualitatively (Moore 2006)	No	Yes: qualitatively (Moore 2006)	Yes: qualitatively (Moore 2006)	No	No	Yes (Moore 2006)
Complex network theory		Yes: topological structure analysis (Albert et al. 2000; Strogatz 2001; Crucitti et al. 2006; Rosato et al. 2007; Zio et al. 2008)	Yes (Dueñas-Osorio et al. 2007; Johansson and Jonsson 2009)	Yes (Motter and Lai 2002; Crucitti et al. 2004; Glass et al. 2004; Kinney et al. 2005)	Yes (Latora and Marchiori 2001; Chassin and Posse 2005; Zio et al. 2008)	Yes (Latora and Marchoiri 2005; Cadini et al. 2009; Hines and Blumsack 2008; Johansson and Jonsson 2009; Zio et al. 2008)	Yes (Latora and Marchoiri 2005; Zio et al. 2008)	No	Yes (Motter 2004; Dueñas-Osorio and Vemuru 2009)
Agent-based modeling and simulation		Yes (Schläpfer et al. 2008)	Yes (Panzieri et al. 2004; Casalicchio et al. 2007)	Yes (Schläpfer et al. 2008)	Yes (Schläpfer et al. 2008)	Yes (Schläpfer et al. 2008)	No	No	No
Dynamic control system theory	Transfer function	Yes (MIA 2010)	Yes (MIA 2010)	No	Yes (MIA 2010)	No	No	No	No

from them. Moreover, the structure of the CI under analysis may be hidden by the fact that the data are often presented in an aggregate form (Dekker 2005; Debon et al. 2010) so that, for example, the propagation of cascading failures may not be properly accounted for. The wealth of statistical models available for the analysis of engineered systems (Lord et al. 2005) can also be a drawback, in that a proper choice must be made of the most suitable model for the specific CI which best fits the physics of the provided service. In this sense, special emphasis must be put on comparing the accuracy and usefulness of the models by means of goodness of fit statistics.

Statistical models of differing complexity have, for example, been suggested in the literature for predicting pipe breaks in water distribution systems, from proportional hazard models (PHMs) to accelerated lifetime models (ALMs), to generalized linear models (GLMs) (Debon et al. 2010). These models were designed to show the impact on the CI of each variable describing the failure of an individual component. The "Cox regression model", also called PHM or duration model, is designed to analyze the time lapse until an event occurs or between events. One or more predictor variables, called covariates, are used to predict a status (event). The PHM has been widely used in analyzing survival data of components, for reliability modeling purposes (Cox 1972). ALMs are designed to analyze what influence the covariates have on the component failure time; they can be viewed as analogous to the PHMs; but, while in the ALMs the covariates act on the time to failure, in the PHMs it is the failure hazard that is affected by the covariates (Kleiner and Rajani 2001). GLMs are an extension of linear models for non-normal distributions of the response variable and non-linear transformations. A regression model constitutes a specification for the variable mean in terms of a small number of unknown parameters corresponding to the covariates. A GLM provides a method for estimating a function of the mean of the response variable as a linear combination of the set of predictive variables (McCullagh and Nelder 1989).

These statistical techniques have been proposed as tools for decision support in the diagnosis and rehabilitation of CIs, e.g. water supply systems (Yamijala et al. 2009), with the additional aim of identifying the systems most critical parameters.

However, criticality ranking of the systems components is not possible by resorting to statistical techniques only, because the data are typically presented in an aggregate form and no identification of the topological structure is accounted for.

4.2 Probabilistic Modeling

This approach encompasses a variety of methods used for the characterization of CIs, such as Markov chains (MCs), Markov/Petri nets (MPNs), probabilistic dynamics modeling and Bayesian networks (BNs).

MCs and MPNs rely on the definition of probabilities of transition of the system components among their reachable states; this may pose significant challenges because of the exponential growth in the number of CI configurations to be evaluated (Iyer et al. 2009). The behavior of a CI is described by its states and by the possible transitions between these states. The system states are defined by the states of the components comprising the system. The components are not restricted to having only two possible states but rather may have a number of different states, such as functioning, in standby, completely failed and under maintenance. The various failure modes of a component may also be defined as states. However, if the system is made by several multi-state components, the number of system states increases rapidly. The state-space representation of the system behavior is, therefore, suitable only for relatively small systems. The possible transitions between the states are illustrated with the aid of a state-space diagram also known as a Markov diagram. The transitions are caused by various mechanisms and activities such as failures, repairs, replacements and switching operations. Complicated repair and switching strategies, common cause failures and other realistic aspects of system behavior can be included in the state-space model. Mathematical modeling of systems suffering from cascading failures propagation has been developed using a continuous-time Markov chain (CTMC) model of a CI operating in a randomly changing environment and as probabilistic cascading failures (Iyer et al. 2009). Probabilistic cascading is considered in the sense that the failure of a component of type *i* causes a component of type *j* to fail simultaneously with a given probability, with all the failures in a cascade being mutually independent. The model includes an environment variable that changes as a CTMC, and this allows us to model randomly changing conditions under which the system operates. The model allows for the evaluation of vulnerability indicators such as the unavailability and mean-time-to-failure (MTTF) of the service provided by the CI.

Probabilistic dynamics models can be considered to overcome the computational limitations of the previous methods; yet, their analysis is affected by the drawback that the identification of the system logical structure is not accounted for (Watts 2002; Dobson et al. 2005). Also, probabilistic dynamics models allow accounting for (inter)dependencies and interactions among several CIs (Newman et al. 2005), while MCs and MPNs have been typically used to analyze isolated systems, or to analyse the (inter)dependencies only for high level qualitative analysis (Sultana and Chen 2009). In the latter case, the actions and the effects of the (inter)dependencies among these CIs are postulated by the analyst at the global CI scale, and do not emerge as the result of the local (inter)dependencies among components of different CIs (Sultana and Chen 2009).

BN analysis is a probabilistic approach that can be used for modeling and predicting the behavior of a system, based on observed stochastic events. A BN is a model that represents the interactions among the components in a system, from a probabilistic perspective. The representation is illustrated via a directed acyclic graph, where the nodes represent the variables and the links between each pair of nodes represent the causal relationships between the variables. From a network reliability perspective, the variables of a BN are defined as the components in the network, while the links represent the interaction of the nodes leading to system "success" or "failure". A fundamental assumption for the construction of a BN is that in general, the strength of the interaction/influence among the graph nodes is uncertain, and thus this uncertainty is represented by assigning a probability of existence to each of the links joining the different nodes. For non-trivial systems, i.e. systems not following a series, parallel or any combination of these configurations, the failure/success probability of a system usually depends on the failure/success of a non-evident collection of components. In a BN this dependency is represented as a directed link between two components, forming a child and parent relationship, so that the dependent component is called the child of the other. Therefore, the success probability of a child node is conditional on the success probabilities associated with each of its parents. The conditional probabilities of the child nodes are calculated by using the Bayes' theorem via the probability values assigned to the parent nodes. Also, the absence of a link between any two nodes of a BN indicates that these components do not interact for system failure/success, thus they are considered independent of each other and their probabilities are calculated separately. Holistic methods have been devised for constructing BN models for estimating the two-terminal reliability of abstract networks (i.e. the probability of a connection between a selected pair of source and target nodes in the network) (Doguc and Ramirez-Marquez 2009). BNs have also been shown useful in assessing the probabilistic relationships and identifying probabilistic mappings among the components of a CI. Methods exist that use historical data about the system to be modeled as a BN and provide for the automated construction of the BN model. In this respect, data mining algorithms are used for finding associations between system components, and thus building the BN model. These algorithms use a heuristic to provide efficient and accurate results while searching for associations.

4.3 Risk Analysis

This approach can be divided into two lines of analysis: the first entails the qualitative assessment of system vulnerabilities by expert judgment and tabular methods (Moore 2006; Piwowar et al. 2009), while the second entails the quantitative part (Apostolakis and Lemon 2005; Flammini et al. 2009), with the further aim of ranking systems elements according to their criticality and of assessing their cascade failure dynamics (Koonce et al. 2008). To a certain extent, the risk analysis approach to the vulnerability of CIs can be considered a general framework of analysis, since it often takes advantage of other approaches and tools, i.e. power flow analysis for electrical transmission networks (Koonce et al. 2008) and network analysis (Apostolakis and Lemon 2005).

Methods such as fault tree (FT) and event tree (ET) analysis as core methods of probabilistic risk assessment (PRA) have been applied to the vulnerability analysis of CIs for protecting the systems against malevolent actions (Piwowar et al. 2009). The approach comprises a step-by-step process typical of PRA : (1) systemic analysis, in which the system itself and its surroundings are analyzed to identify all the parameters that could interact with it; (2) analysis of the interactions between aggressors' profiles and systems, in which the systems are ranked according to the degree of risk of being attacked; (3) assessment of vulnerabilities and determination of key points of vulnerability, in which a panel of subject matter experts and a PRA process normative expert conduct a study and weigh the importance of the system elements to qualify them according to several criteria concerning the attack; (4) building of scenarios, taking into account system functionalities, the global environment, and the geopolitical context; (5) updating the security systems, considering the results of the previous steps.

Yet, the approach presents some inconveniences related to the size of the distributed system to analyze, the number of scenarios of attack to be considered, the human subjectivity as a base for all the quantifications to assess vulnerabilities on a given CI, and the needs to be updated frequently, because of the evolving geopolitics and the introduction of new factors in the whole system (a new apparel, new ways of productivity, new entrance or exit points, etc.).

A risk-based quantitative method for the identification and prioritization of vulnerabilities in interdependent CIs has been proposed in (Apostolakis and Lemon 2005). The CI is modeled as interconnected digraphs and the graph theory is employed to identify the candidate vulnerable scenarios. These scenarios are screened for the susceptibility of their elements to a terrorist attack, and a prioritized list of elements vulnerabilities is produced. The prioritization methodology is based on multi-attribute utility theory (Morgan et al. 2000). The impact of losing infrastructure services is evaluated using a value tree that reflects the perceptions and values of the decision maker and the relevant stakeholders. These results, which are conditional on a specified threat, are provided to the decision maker for use in risk management. Interestingly, this method embeds the vulnerability quantification into the framework of the stakeholders' perspective, making it suitable for a realistic ranking of vulnerabilities in CIs.

To account for cascading failure propagation in CIs, the previous method has been extended and applied to the risk analysis of a bulk power system (Apostolakis and Lemon 2005). The analysis was complemented with a power flow simulation model to determine the likelihood and extent of power outages, when components within the system fail to perform their designed functions due to both random causes and malevolent acts. The consequences associated with these failures are determined by looking at the type and number of customers affected; stakeholder input is used to evaluate the relative importance of these consequences. The methodology provides a ranking of each system component by its risk significance to the stakeholders.

4.4 Complex Network Theory

Complex network theory methods can be applied to the analysis of CIs for (a) helping to identify preliminary vulnerabilities by topology-driven and dynamical analyses and (b) guiding and focusing further detailed analyses of critical areas.

The interconnection structure of a CI can be represented by an unweighted network, where the edges (arcs) between nodes are either present or not. The topological analysis of the corresponding representative graphs has received a renewed enthusiasm from the works of Strogatz and Barabási, who pointed at the presence of a "selective pressure" able to determine a common growth mechanism in a number of diverse evolving complex systems (from those representing technological systems to those related to biological and social networks; Watts and Strogatz 1998; Jeong et al. 2001). Such a growth mechanism determines the development of structures toward a class of mathematical graphs labeled as "scale-free". It has been shown that those network structures possess considerable robustness against random faults but large vulnerability to deliberate attacks, with respect to randomly grown networks (Albert et al. 2000). On the practical side, it has also been shown that, quite often, large technological infrastructures do not show a clear scale-free structure, mainly due to technological constraints which limit the arbitrary growth of the nodes degrees (i.e. the number of connections pointing to the nodes) (Amaral et al. 2000).

Topological analysis based on classical graph theory can unveil relevant properties of the structure of a network system (Albert et al. 2000; Strogatz 2001) by (i) highlighting the role played by its components (nodes and connecting arcs) (Crucitti et al. 2006; Zio et al. 2008), (ii) making preliminary vulnerability assessments based on the simulation of faults (mainly represented by the removal of nodes and arcs) and the subsequent re-evaluation of the network topological properties (Rosato et al. 2007; Zio et al. 2008). In a topological analysis, a CI is represented by a graph $G(N, K)$, in which its physical constituents (components) are mapped into N nodes connected by K unweighted edges, representing the links of physical connection among them. The focus of topological analysis is on the structural properties of the graphs on the global and local scales, e.g., as represented by their characteristic path length, L, defined as the number of arcs in the shortest path between two nodes averaged over all pairs of nodes, and average clustering coefficient, C, a measure of the extent to which nodes tend to form small groups (Watts and Strogatz 1998). Average global measures, such as L, give indications on the extent to which each node in the system is connected with any other nodes, while average local measures, like C, assess to what extent the first neighbors of each node are connected among each other. In spite of the usefulness of the topological analysis of the unweighted network of a CI and of the insights it provides, empirical results show that it cannot capture all rich and complex properties observed in real CIs, so that there is a need for extending the models beyond pure unweighted, structural topology (Boccaletti et al. 2006; Eusgeld et al. 2009).

Indeed, along with a complex topological structure, many real networks display a marked physical heterogeneity in the capacity and intensity of the connections. Examples are different impedance and reliability characteristics of overhead lines in electrical transmission networks (Hines and Blumsack 2008; Eusgeld et al. 2009), unequal traffic on roads which affects accident probability (Zio et al. 2008) or different routing capacities of the Internet links (Latora and Marchoiri 2005). To describe such heterogeneities numerical weights can be assigned to each link of the representative network, measuring the "strength" of the connection. In this way, the functional behavior of the CI is somewhat embedded into a generalized, but still simple, topological analysis framework. Global and local measures can then be introduced for the statistical characterization of weighted networks (Latora and Marchiori 2001).

Further, in real network systems, another important dimension to add to the vulnerability characterization is the dynamics of flow of the physical quantities in the network. From the analysis point of view, this entails considering the interplay between structural characteristics and dynamical aspects, which makes the modeling very complicated since the load and capacity of each component, and the flow through the network are often highly variable quantities both in space and time. Functional models have been developed to capture the basic realistic features of CI networks within a weighted topological analysis framework, i.e. disregarding the representation of the individual dynamics of the CIs elements. These models have shed light on the way complex networks react to faults and attacks, evaluating their consequences when the dynamics of flow of the physical quantities in the network is taken into account. The response behavior often results in a dramatic cascade phenomenon due to avalanches of node breakings (Motter and Lai 2002; Motter 2004; Zio and Sansavini 2008). Abstract modeling paradigms for analyzing the system response to cascading failures have been introduced to guide successive detailed simulation focused on the most relevant physical processes and network components (Motter and Lai 2002; Motter 2004; Dobson et al. 2005; Zio and Sansavini 2008). Despite their apparent simplicity, these models provide indications on the elements criticality for the propagation process (Zio and Sansavini 2011a) and on the actions that can be performed in order to prevent or mitigate the undesired effects (Motter 2004). It is observed that improvements in network components alone do not ensure system robustness or protection against disproportionate cascading failures (Dueñas-Osorio and Vemuru 2009) instead, topological changes are needed to increase cascading robustness at realistic tolerance levels. Moreover, regardless of the nature of the events triggering the failure cascade, the additional loss of performance due to cascading failures can be orders of magnitude larger than the initial loss of performance.

Finally, the complex network theory models allow accounting for (inter)dependencies among different CIs, to assess the influences and limitations which interacting infrastructures impose on the individual systems operating conditions, for avoiding fault propagation by designing redundancies and alternative modes of operations, for detecting and recognizing threats (Zimmerman 2001; Dueñas-Osorio et al. 2007; Johansson and Jonsson 2009). In developing modeling and simulation frameworks that allow the coupling of interdependent infrastructures, it is important to know that simply linking existing infrastructure models together fails to capture the emergent behavior arising in interdependent CIs (Zio and Sansavini 2011b).

4.5 Agent-Based Modeling and Simulation

Agent-based modeling (ABM) has been shown to offer an attractive modeling paradigm for describing the dynamic system operational behavior, with close adherence to the reality of the coupled processes involved (D'Inverno and Luck 2004). One of the major advantages of ABM for simulating CIs is the possibility to include physical laws into the simulation and to emulate the behavior of the infrastructure as it emerges from the behaviors of the individual agent and their interactions. In other words, the overall system behavior results from the interactions among the multiple single agents of different kinds which make up the system (Schläpfer et al. 2008). This modeling approach integrates the spectrum of different phenomena and mechanisms which may occur, thus generating a multitude of representative time-dependent event chains.

In order to include technical and non-technical factors and physical laws in the vulnerability assessment of a critical infrastructure, a two-layer approach can be deployed (Schläpfer et al. 2008). For example, an electric power system can be thought of as a stochastic hybrid system that can be modeled by a finite state machine (FSM) whose states involve continuous variables with uncertain dynamics; transitions in this machine correspond to outages of generation and transmission equipment. The conceptual modeling framework consists in the abstraction of the relevant components of the system as individual interacting agents. Agents are used to model both technical components (such as the electric power generators) and non-technical components (such as grid operators). The different agents interact with each other directly (e.g. generator dispatch) or indirectly, e.g. via the physical network. Each agent is modeled by attributes and rules of behavior. An example for an attribute is a technical component constraint such as the rating of a transmission line in the electric power system. The rules of behavior are represented by using FSM and include both deterministic and stochastic time-dependent, discrete events. A deterministic event is, for instance, the outage of a component when reaching a failure threshold, while stochastic processes are probabilistic component failure models simulated by Monte Carlo techniques (Billington and Li 1994; Marseguerra and Zio 2002).

ABM was used to assess the availability of bulk power systems for mid- and short-period power system planning purposes (Schläpfer et al. 2008). It was shown that the probability of cascading failures is subject to phase transitions and abrupt changes that result from only small changes in the system stress. The level of modeling detail offered by ABM allows analyzing a multitude of time-dependent availability aspects. Besides technical failures, other factors, such as natural hazards, institutional weaknesses or security-related issues can be integrated. The main problems are related to the slow simulation speed and the large number of parameters to be input in the analysis. However, by focusing on specific safety aspects, the model can be simplified and the computational burden reduced.

ABM is also a suitable approach to simulate interdependencies among CIs (Panzieri et al. 2004) by introducing interconnected agents: i.e., independent systems that autonomously elaborate information and resources in order to define their outputs; the latter become inputs for other agents, and so on (Luck et al. 2003). This approach is particularly useful for situations, as is the case of infrastructure interdependencies, with sparse or non-existent macro scale information; ABM is able to use the rich sources of micro-level data to develop interaction forecasts. One disadvantage of these simulation models is that the complexity of the computer programs tends to obscure the underlying assumptions and the inevitable subjective inputs. Another disadvantage is the difficulty to acquire detailed information about each single infrastructure. This task appears, by its own, a difficult challenge, because this kind of information is considered very sensible by infrastructure stakeholders due to the relevance for their business. In Hopkinson et al. (2003), to partially overcome this problem, commercial simulators are integrated into a message-broker framework.

4.6 Dynamic Control System Theory

Dynamic control system theory can be applied as a quantitative analytical method to assess the (inter)dependencies among CIs using the transfer functions and the corresponding frequency responses (MIA 2010). In order to apply this method, studied infrastructure(s) need to be assumed as a linear time invariant (LTI) dynamic system.

The input/output relationship between infrastructures is able to be represented and quantified by deriving the transfer functions in the domain of Laplace (s) or the domain of frequency ($j\omega$). Mason's formula, a method determining the control function of a given control loop after identifying its corresponding signal flow graph, can be applied to define all the required transfer functions such as the transfer function for a component of one infrastructure or overall global transfer function. It should be noted that the overall transfer function is constituted by the composition of the transfer functions of each component. A final possible representation of the transfer function is in the form of poles/zeros:

$$G(s) = \frac{\mu \times \prod_j (s - z_j)}{s^g \times \prod_i (s - p_i)} \tag{4.1}$$

where j = the number of zeros, i = the number of poles (MIA 2010).

Based on the overall transfer function and its corresponding frequency responses, the stability of interdependencies between infrastructures can be evaluated using BODE and NYQUIST diagrams (MIA 2010).

The approach of the dynamic control system theory is a novel method, which brings the classic control system theory to the area of CI protection. Instead of using the time domain, two alternative domains (domain of Laplace and frequency) are used for the purpose of evaluation and assessment. The development of transfer functions is the most essential part of this approach and has strong influences on the accuracy of the final results, which could be further complicated due to the complexities of the studied infrastructures. Another drawback of this approach lies in the fact that hidden vulnerabilities caused by (inter)dependencies are not able to be estimated since all the links between studied infrastructures have been determined during the transfer function development. The applicability/feasibility of this approach is still under discussion and needs to be further proved.

References

Albert R, Jeong H, Barabási A-L (2000) Error and attack tolerance of complex networks. Nature 406:378–382
Amaral LAN, Scala A, Barthélémy M, Stanley HE (2000) Classes of small-world networks. Proc Natl Acad Sci USA 97:11149–11152
Angel M, Werley A-K (2007) Stochastic model for power grid dynamics. In: Proceedings of the 40th Hawaii international conference on system sciences. January 3–6, 2007, Big Island, Hawaii
Apostolakis E-G, Lemon M-D (2005) A screening methodology for the identification and ranking of infrastructure vulnerabilities due to terrorism. Risk Anal 25(2):361–376
Augutis J, Krikštolaitis R, Šidlauskas K, Martišauskas L, Matuziene V (2010) Modeling of energy supply disturbances in network systems. In: Briš R, Guedes Soares C, Martorell S (eds) Reliability, risk and safety: theory and applications. Taylor and Francis, London
Billington R, Li W (1994) Reliability assessment of electric power systems using Monte Carlo methods. Plenum Press, New York
Boccaletti S, Latora V, Moreno Y, Chavez M, Hwang D-U (2006) Complex networks: structure and dynamics. Phys Rep 424:175–308
Buzna L, Peters K, Ammoser H, Kühnert C, Helbing D (2007) Efficient response to cascading disaster spreading. Phys Rev E 75(5):056107
Cadini F, Zio E, Petrescu C-A (2009) Using centrality measures to rank the importance of the components of a complex network infrastructure. In: Proceedings of CRITIS'08, 13–15 October 2008, Rome, Italy, pp 155–167
Casalicchio E, Galli E, Tucci S (2007) Federated agent-based modeling and simulation approach to study interdependencies in IT critical infrastructures. In: Proceedings of the 11th IEEE symposium on distributed simulation and real-time applications, Chania, Crete Island, Greece
Chassin P-D, Posse C (2005) Evaluating North American electric grid reliability using the Barabasi–Albert network model. Physica A 355:667–677
Chen J, Thorp S-J, Dobson I (2005) Cascading dynamics and mitigation assessment in power system disturbances via a hidden failure model. Int J Electr Power Energ Syst 27:318–326
Coffman E-G Jr, Ge Z, Misra V, Towsley D (2002) Network resilience: exploring cascading failures within BGP. In: Proceedings of the 40th annual Allerton conference on communications, computing and control, Monticello, Illinois, USA
Cox D (1972) Regression models and life tables (with discussion). J R Stat Soc Ser B 34(2): 187–220
Crucitti P, Latora V, Marchiori M (2004) Model for cascading failures in complex networks. Phys Rev E 69:045104(R)
Crucitti P, Latora V, Porta S (2006) Centrality in networks of urban streets. Chaos 16(1–9): 015113
D'Inverno M, Luck M (2004) Understanding agent systems. Springer, Berlin
Debon A, Carrion A, Cabrera E, Solano H (2010) Comparing risk of failure models in water supply networks using ROC curves. Reliab Eng Syst Saf 95:43–48
Dekker AH (2005) Simulating network robustness for critical infrastructure networks, conferences in research and practice in information technology. In: Estivill-Castro V (ed) Proceedings of the 28th Australasian computer science conference, the University of Newcastle, vol 38. Newcastle, Australia
Dobson I, Carreras BA, Lynch V, Newman DE (2004) Complex systems analysis of series of blackouts: cascading failure, criticality and self-organization, bulk power system dynamics and control—VI. Cortina d'Ampezzo, Italy, pp 438–451
Dobson I, Carreras BA, Newman DE (2005) A loading-dependent model of probabilistic cascading failure. Prob Eng Inform Sci 19:15–32
Doguc O, Ramirez-Marquez EJ (2009) A generic method for estimating system reliability using Bayesian networks. Reliab Eng Syst Saf 94:542–550

Dueñas-Osorio L, Vemuru S-M (2009) Cascading failures in complex infrastructure systems. Struct Saf 31:157–167

Dueñas-Osorio L, Craig IJ, Goodno JB, Bostrom A (2007) Interdependent response of networked systems. J Infrastruct Syst 13(3):185–194

Eusgeld I, Kröger W, Sansavini G, Schläpfer M, Zio E (2009) The role of network theory and object-oriented modeling within a framework for the vulnerability analysis of critical infrastructures. Reliab Eng Syst Saf 94(5):954–963

Flammini F, Gaglione A, Mazzocca N, Pragliola C (2009) Quantitative security risk assessment and management for railway transportation infrastructures. In: Proceedings of critical information infrastructure security, third international workshop, CRITIS 2008, Rome, Italy, October 13–15, 2008. Revised papers, LNCS, Vol. 5508. Springer-Verlag, Berlin, Heidelberg, pp 180–189

Glass JR, Beyeler EW, Stamber LK (2004) Advanced simulation for analysis of critical infrastructure: Abstract cascades, the electric power grid, and fedwire 1 (SNL paper SAND 2004-4239). Albuquerque, New Mexico 87185 and Livermore, California 94550

Guida M, Longo M, Postiglione F (2010) Reliability analysis of next generation mobile networks. In: Briš R, Guedes Soares C, Martorell S (eds) Reliability, risk and safety: theory and applications. Taylor and Francis, London

Hines P, Blumsack S (2008) A centrality measure for electrical networks. In: Proceedings of the 41st Hawaii international conference on system science, Big Island, Hawaii

Hopkinson K, Birman K, Giovanini R, Coury D, Wang X, Thorp J (2003) EPOCHS: integrated commercial off-the-shelf software for agent-based electric power and communication simulation. In: Proceedings of the 2003 winter simulation conference, New Orleans, LA, 7–10 December 2003, pp 1158–1166

Iyer MS, Nakayama KM, Gerbessiotis VA (2009) A Markovian dependability model with cascading failures. IEEE Trans Comput 58(9):1238–1249

Jeong H, Mason SP, Barabasi A-L, Oltvai ZN (2001) Lethality and centrality in protein networks. Nature 411:41–42

Johansson J, Jonsson H (2009) A model for vulnerability analysis of interdependent infrastructure networks. In: Martorell et al. (eds) Safety, reliability and risk analysis: theory, methods and applications. Proceedings of ESREL 2008 and 17th SRA Europe annual conference, 22–25 September 2008, Valencia, Spain, Taylor & Francis Group, London

Kinney R, Crucitti P, Albert R, Latora V (2005) Modeling cascading failures in the North American power grid. Eur Phys J B 46:101–107

Kleiner Y, Rajani B (2001) Comprehensive review of structural deterioration of water mains: statistical models. Urban Water 3(3):131–150

Koonce AM, Apostolakis GE, Cook BK (2008) Bulk power risk analysis: ranking infrastructure elements according to their risk significance. Int J Electr Power Energy Syst 30:169–183

Krings A, Oman P (2002) A simple GSPN for modeling common mode failures in critical infrastructures. In: Proceedings of the 36th annual Hawaii international conference on system sciences (HICSS'03), Big Island, Hawaii

Langeron Y, Barros A, Grall A, Bérenguer C (2010) Reliability assessment of network-based safety-related systems. In: Briš R, Guedes Soares C, Martorell S (eds) Reliability, risk and safety: theory and applications. Taylor and Francis, London

Laprie J-C, Kanoun K, Kaâniche M (2007) Modelling interdependencies between the electricity and information infrastructures. In: Proceedings of the 26th international conference on computer safety, reliability, and security (SAFECOMP 2007), Nuremberg, Germany, LNCS 4680/2009

Latora V, Marchiori M (2001) Efficient behavior of small-world networks. Phys Rev Lett 87(19):198701 (1–4)

Latora V, Marchoiri M (2005) Vulnerability and protection of infrastructure networks. Phys Rev E 71:015103 (1–4)

Lord D, Washington PS, Ivan NJ (2005) Poisson, Poisson-gamma and zero inflated regression models of motor vehicle crashes: balancing statistical fit and theory. Accid Anal Prev 37:35–46

Luck M, McBurney P, Preist C (2003) Agent technology: enabling next generation computing (A roadmap for agent based computing). AgentLink II. University of Southampton, Southampton, UK

Marseguerra M, Zio E (2002) Basics of the Monte Carlo method with application to system reliability. LiLoLe-Verlag GmbH, Hagen, Germany

McCullagh P, Nelder J (1989) Generalized linear models. Chapman & Hall, London

MIA (2010) Definition of a methodology for the assessment of mutual interdependencies between ICT and electricity generation/transmission infrastructures. Final report, September 2010, Italian National Agency for New Technology, Energy and Environment, Italy

Moore AD (2006) Application of the API/NPRA SVA methodology to transportation security issues. J Hazard Mater 130:107–121

Morgan MG, Florig HK, DeKay ML, Fischbeck P (2000) Categorizing risks for risk ranking. Risk Anal 20:49–58

Motter A-E (2004) Cascade control and defense in complex networks. Phys Rev Lett 93(9): 098701(1-4)

Motter A-E, Lai Y-C (2002) Cascade-based attacks on complex networks. Phys Rev E 66(1–4): 065102

Newman D-E, Nkei B, Carreras BA, Dobson I, Lynch VE, Gradney P (2005) Risk assessment in complex interacting infrastructure systems. In: Proceedings of the 38th Hawaii international conference on system sciences, Big Island, Hawaii

Panzieri S, Setolaand R, Ulivi G (2004) An agent based simulator for critical interdependent infrastructures. In: Proceedings of the 2nd international conference on critical infrastructures CRIS2004: October 25–27, 2004, Grenoble, France

Paté-Cornell ME, Guikema SD (2002) Probabilistic modeling of terrorist threats: a systems analysis approach to setting priorities among countermeasures. Mil Oper Res 7(3):5–23

Piwowar J, Chatelet E, Laclemence P (2009) An efficient process to reduce infrastructure vulnerabilities facing malevolence. Reliab Eng Syst Saf 94:1869–1877

Rosato V, Bologna S, Tiriticco F (2007) Topological properties of high-voltage electrical transmission networks. Electr Pow Syst Res 77:99–105

Schläpfer M, Kessler T, Kröger W (2008) Reliability analysis of electric power systems using an object-oriented hybrid modeling approach. In: Proceedings of the 16th power systems computation conference, Glasgow

Strogatz SH (2001) Exploring complex networks. Nature 410:268–276

Sultana S, Chen Z (2009) Modeling flood induced interdependencies among hydroelectricity generating infrastructures. J Environ Manage 90:3272–3282

Watts D-J (2002) A simple model of global cascades on random networks. Proc Natl Acad Sci USA 99(9):5766–5771

Watts D-J, Strogatz SH (1998) Collective dynamics of 'small-world' networks. Nature 39: 440–442

Yamijala S, Guikema DS, Brumbelow K (2009) Statistical models for the analysis of water distribution system pipe break data. Reliab Eng Syst Saf 94:282–293

Zimmerman R (2001) Social implications of infrastructure network interactions. J Urban Technol 8(3):97–119

Zio E, Sansavini G (2008) Modeling failure cascades in networks systems due to distributed random disturbances and targeted intentional attacks. In: Martorell et al. (eds) Safety, reliability and risk analysis: theory, methods and applications. Proceedings of ESREL 2008 and 17th SRA Europe annual conference, Valencia, Spain, Taylor & Francis Group, London, 22–25 September 2008,

Zio E, Sansavini G (2011a) Component criticality in failure cascade processes of network systems. Risk Anal. doi: 10.1111/j.1539-6924.2011.01584.x

Zio E, Sansavini G (2011b) Modeling interdependent network systems for identifying cascade-safe operating margins. IEEE Trans Reliab 60(1):94–101

Zio E, Sansavini G, Maja R, Marchionni G (2008) An analytical approach to the safety of road networks. Int J Reliab Qual Saf Eng 15(1):67–76

Chapter 5
Conceptual Frameworks for Vulnerability Assessment

Critical Infrastructures (CIs) have basic traits in common such as large-size, wide-area coverage, complexity and interconnectedness but show significant differences in detail, even when limited to physical-engineered CIs. The challenges for understanding, modeling and analyzing these systems with regard to vulnerability are immense and far away from being resolved. Approaches can be of two types, empirical and predictive, or a combination of both, as they complement each other (Johansson and Hassel 2010). By empirical investigations, previous events are studied in order to understand the behavior of CIs and the (inter)dependencies between them as well as to identify patterns, while predictive approaches aim to model the behavior of single or a group of CIs to find potential high consequence cases and non-obvious vulnerabilities.

5.1 General Outline

Applied methods, in particular for predictive analysis, have different view points, e.g., functional or structural, different levels of abstraction, different focus, objectives and metrics, different degrees of maturity and based on different levels of available information and knowledge, etc., None of the available methods alone is capable to address and tackle the variety of issues of vulnerability assessment including the impact of (inter)dependencies. It is commonly agreed that a universal, all-encompassing ("silver-bullet") approach or model which accounts for all issues does not exist (Eusgeld et al. 2009; Johansson and Hassel 2010).

Therefore, a conceptual framework is needed to bring all aspects, attributes and capabilities together and which goes beyond the pure analytical framework (architecture). Attempts have been made to develop conceptual frameworks with varying degrees of precision (see also Aven 2010) and focus, e.g., modeling interdependencies (Dudenhoeffer et al. 2006; Johansson and Hassel 2010), covering safety and security (Aven 2010) or cases of malevolence (Piwowar et al. 2009).

W. Kröger and E. Zio, *Vulnerable Systems*, DOI: 10.1007/978-0-85729-655-9_5,

Some proposals address single sectors like the Internet infrastructure (Qu et al. 2002), electric power grids and control systems (MIA 2010) or physical-engineered, (inter)dependent CIs in general (Eusgeld et al. 2009; Utne et al. 2009).

The goal of vulnerability analysis as outlined in Chap. 1 can be translated into a set of generic questions to be answered such as:

- What are the end states of interest for the given system(s) and how to define their boundaries?
- What are the threats and hazards of relevance that the system(s) under consideration might be exposed to?
- What is the sensitivity (susceptibility) and resilience of the system(s) experiencing them?
- What are resulting cascades and which (inter)dependencies can be identified? What is their impact and which are the high consequence scenarios?
- What are the uncertainties involved?

And finally:

- What are the obvious and non-obvious ("hidden") vulnerabilities and how can they be reduced and/or better managed?

Further specification of these questions is necessary when starting the vulnerability assessment of a specific system or a set of interconnected systems, and this can only be done in close cooperation between the analysts and the orderer ("problem owner") or recipient of the analysis, e.g., a public authority or private entity. The questions of what vulnerability/risk is and how it can be measured in the specific case, and at which level of accuracy, need to be answered precisely.

5.2 Outline of Stepwise Framework Approaches

It seems to be commonly agreed that a conceptual framework for vulnerability analysis should follow a stepwise, problem-driven approach tailored to the needs of the analysis. Two examples are given here, based on two different starting viewpoints.

5.2.1 Framework Tailored to Previously Defined, Undesired Events

The first example concerns a framework focused on the analysis of interdependency issues among CIs (Utne et al. 2009) and assumes that undesired events affecting several CIs can be derived from experience or gained by "brain power", and can be selected for further analysis. This so-called DECRIS conceptual framework for cross sector infrastructure vulnerability analysis starts with a

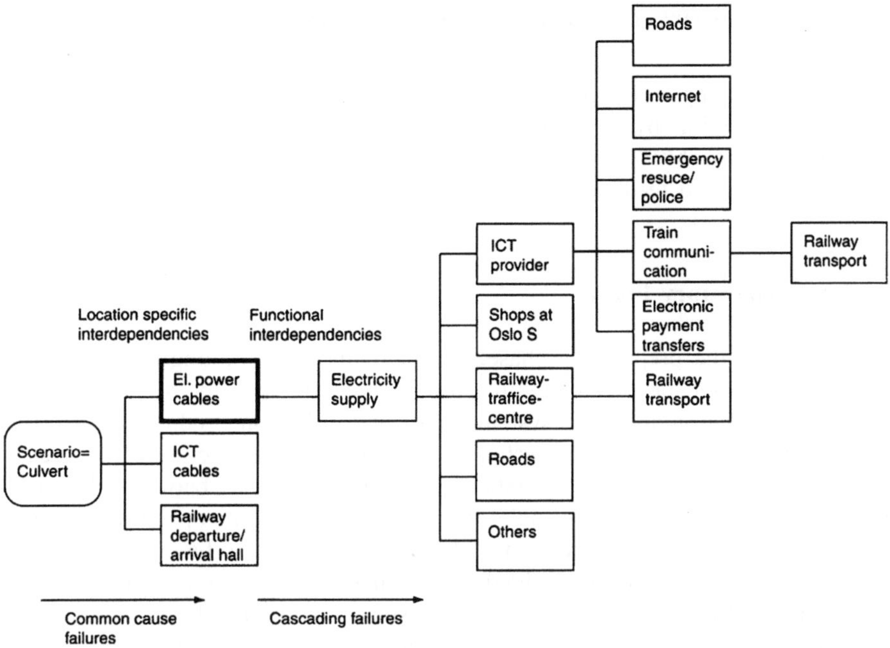

Fig. 5.1 First and second order functional interdependencies identified by applying the DECRIS framework (Utne et al. 2009). ("STOP" box indicates that we choose not to further analyze this object or function in detail)

screening analysis of experienced undesired events (phase 1) to select events, involving several infrastructures of interest, for further detailed analysis (phase 2). The first task of phase 2 aims at developing those selected, undesired events into accident scenarios and investigating interdependencies between the causes of undesired events and between infrastructures; the development of an adequate system understanding is regarded as an important pre-requisite (sub-task). The second task aims at identifying interdependencies of first or a higher order including location-specific (geo-spatial) and functional ones as main causes of cascading failures. The third task aims at analyzing the risks related to the interdependencies by means of qualitative or semi-quantitative methods using probability/frequency, extent and duration as quantitative measures of an accident scenario. If the semi-quantitative does not provide sufficient information, a quantitative analysis should be carried out (task 4) using network models and fault/event trees. Finally (task 5), risk reducing measures have to be evaluated and the question to which extent the scenario could occur in other locations be answered.

The DECRIS framework has been applied to "damage to culvert/joint conduit of electric power, ICT, and water mains based on an undesired event that occurred in Oslo on November 27, 2007. Figure 5.1 shows first and second order identified functional interdependencies.

5.2.2 Framework Including Scenario Generation

The second example (Eusgeld and Kröger 2008; Eusgeld et al. 2009), includes scenario generation by simulation and, where appropriate, a holistic "system-of-system approach". The proposed framework distinguishes five steps, several decision points and feedback loops (see Fig. 5.2)

5.2.2.1 Preparatory Phase

The first step, the preparatory phase, integrates the task framing and definitions (step 1.1) into the process of familiarization with the system, so as to provide suitable information about the system design and operation conditions. Basically, the preparatory phase aims at reaching a clear definition of the terms involved in the assessment and mutual understanding of the objectives between all parties which have a stake in the analysis of the infrastructure and its operation. It is also important to decide on the spectrum of hazards and threats to be included into the analysis which can be of many-sided nature, including intentional malicious acts. Furthermore, it is necessary to deeply understand failure models and effects of/on each of the components of the CI. For a more effective screening of the system vulnerability (and to find obvious vulnerabilities as early and efficiently as possible) some reasonable simplifications should be made. For example the questions whether the interconnected systems can be "decoupled" and analyzed separately, whether it is acceptable to focus on some weak points first, etc., have to be pre-assessed. These assumptions need re-visiting in a later phase of the assessment (step 3.2).

After key information is collected and pre-assessment is done, the knowledge base should be checked with respect to the availability of methods suitable for the defined tasks. To build adequate capability of analysis, interaction with the scientific community needs to be established where appropriate (*double arrow* in Fig. 5.2).

5.2.2.2 Development of System Understanding

In general, the analyst should have basic know-how about the system (design/layout and related standards, functionality, support and control subsystems, operating modes and conditions, etc.) and about the modeling and analysis methods (including limitations/simplifications). Collection and pre-processing statistical data, in particular about system failures, breakdowns, and near-misses to infer preliminary knowledge should be an essential part of the preparatory phase. This also helps to check whether the system is sufficiently well understood and to sharpen the view on vulnerabilities and (inter)dependencies in particular. Cooperation with and support of system owners/operators must be established to

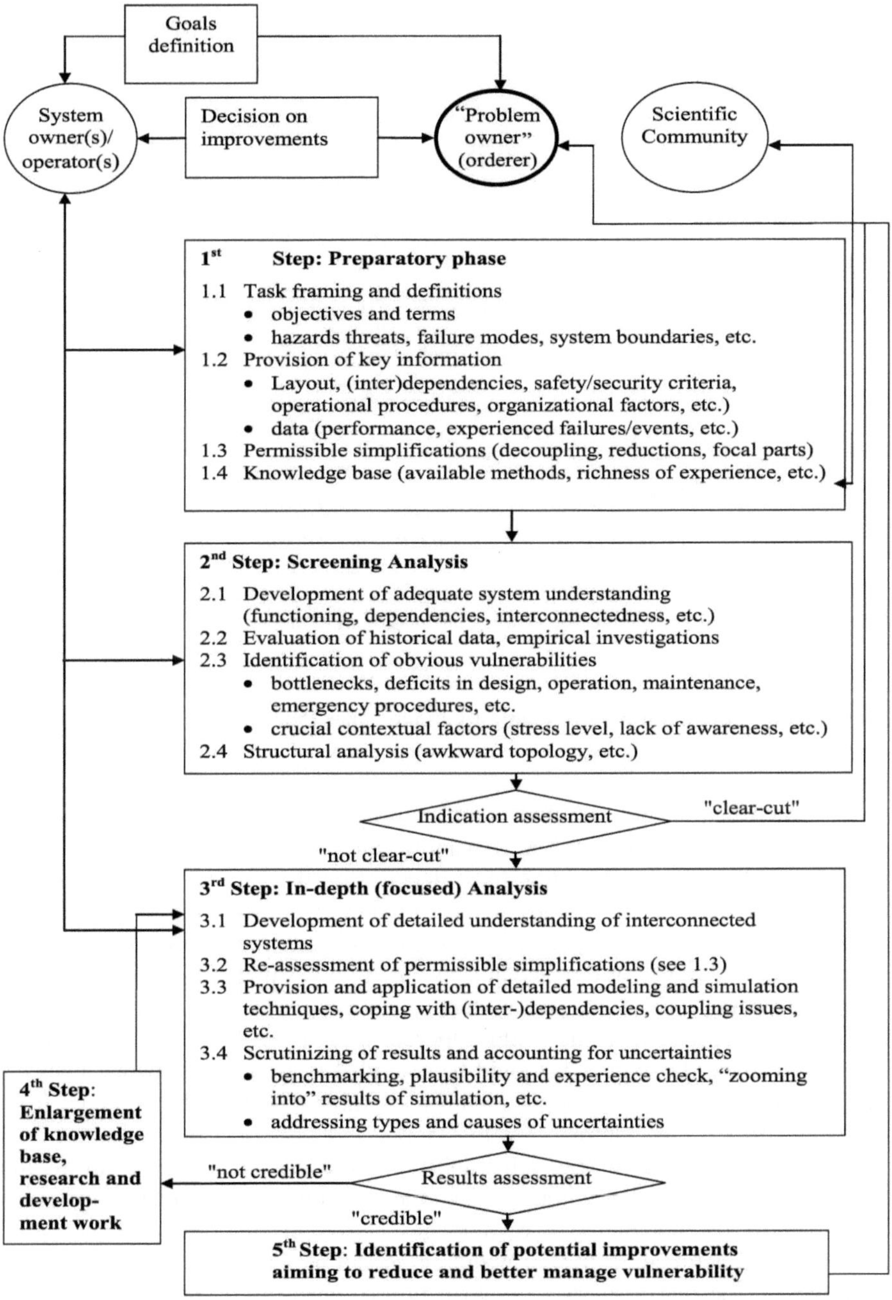

Fig. 5.2 Conceptual framework for the vulnerability analysis of interconnected infrastructures (flow chart-type of illustration; *double arrows* represent two-way interactions)

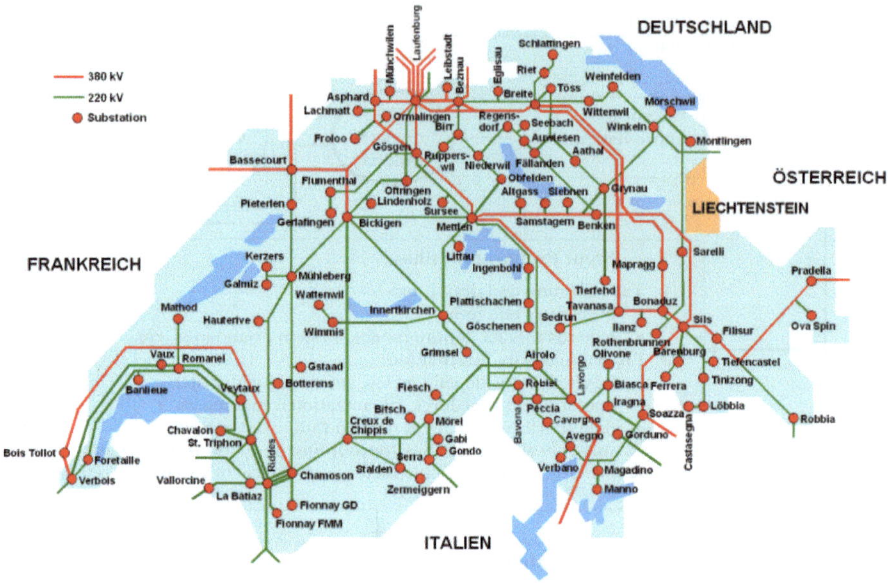

Fig. 5.3 Swiss high voltage electric power transmission system (www.swissgrid.ch)

allocate the information needs and to make sure that the system model is in accordance with reality. System visits ("walk through", "talk through") may be essential.

To which degree the rich and complex properties of the CI being assessed, including (inter)dependencies and cascading failures, need to be understood and how system boundaries should be fixed depends on the purpose and goals of the analysis. Let us take the domestic electric power transmission system as the example for illustration. If you are asked to assess the robustness of the system against random failures or targeted attacks, i.e., to investigate the topology, as part of a "screening analysis", the system boundary is narrow and topological unweighted network analysis seems appropriate; if you want to learn about inhomogeneities, the approach should be extended to weighted network analysis. The information necessary for such analysis is limited and can be procured from open sources, e.g., the Internet (see Fig. 5.3).

On the other hand, if the vulnerability of the electric power transmission system against "major blackouts", i.e., the unavailability of electricity for a large part of the population over a longer period of time, is to be assessed, methods of functional analysis, including modeling the dynamics of physical quantities of the real network (e.g., load flows), need to be applied (e.g., a two layers object-oriented modeling approach). For this "in-depth analysis (step 3 according to the framework), information about the system goes beyond open sources and must be provided by the system operator, with the usual care of the

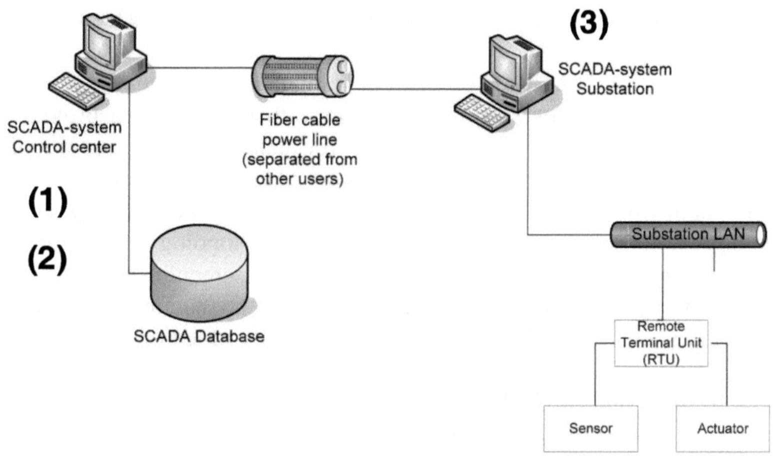

Fig. 5.4 Supervisory control and data acquisition (SCADA) system (simplified Swiss system) (1) Dedicated data exchange between power stations and network operator via PIA (system for exchange among partners) (2) Trading/office systems separated from SCADA (3) Substations have their own guidance systems and can, if needed, be operated by telephone lines; protection systems work independently from SCADA system

confidentiality issues involved; initial load flow conditions are an example of information needed from the system operation. The narrow domestic border cutting-off neighboring countries within the synchronized grid needs to be revisited.

Going one step further and making the disclosure of "hidden" vulnerabilities within the system, or of respective elements, calls for further details. The SCADA systems using modern information and communication technology (ICT) may introduce risks of cyber attacks or common cause failures. The search for potential entry points for potential "outside hackers" requires very detailed information about the layout and operation of this important industrial data acquisition and control system, which can only be provided by the transmission system operator (TSO) and calls for an excellent partnership and trust. Figure 5.4 attempts to illustrate the level of information needed in this case.

5.2.2.3 Screening and In-Depth Analysis

It is assumed that the vulnerability analysis should evolve in two steps where appropriate. A screening-type of analysis might be efficient and sufficient to identify eye-catching, obvious weak points vulnerabilities, e.g., awkward topology, spatial proximity of interconnected systems or bottlenecks, and further actions should focus on eliminating or reducing them. The screening analysis could also prepare the ground for and give steer to the in-depth analysis which may turn out necessary.

The *screening analysis* leads off with development of adequate system understanding (see also previous section); we assume that information provided from system owners/operators (step 1.2) allows for general understanding of main functionalities, states of relevance, interfaces, inter-dependencies, etc. In this phase, the main emphasis is placed on experts' opinions, brainstorming, etc., rather than on application of detailed models. Nevertheless, topology-driven analysis of vulnerabilities may provide an essential support to the screening analysis aiming at identifying the system connection patterns, shortest connection paths, local and global specifics, etc. The techniques used are typically based on complex network theory (see Sect. 6.2).

If the results and insights (indication assessment in Fig. 5.2) gained by screening analysis are not satisfying (not "clear-cut") and major hidden vulnerabilities are still feared, a more sophisticated *in-depth analysis* (step 3) has to be launched.

To achieve a higher degree of accuracy in the vulnerability evaluation, system understanding has to be further developed on the basis of additional information about the system and its operating environment (arrow to and from the owners/operators). Special attention should be placed on (inter)dependencies within or among systems. The re-assessment of simplifications made earlier (including bringing together "decoupled" systems) may call for more sophisticated methods of analysis, e.g., following a system or even a system-of-systems approach. In order to integrate a comprehensive spectrum of different phenomena, stochastic, time-dependent event chains and agent-based modeling approaches combined with Monte Carlo simulation and, where necessary, physical models may be considered (see Sect. 6.5).

Other frameworks exist for this part of the vulnerability analysis, proposing different architectures of analysis for specific types of infrastructure, e.g., for the Internet (see Fig. 5.5 for illustration).

The last but not least step of the in-depth analysis should focus on scrutinizing the results obtained and accounting for the uncertainties of both aleatory (stochastic) and epistemic (lack of knowledge including data and modeling) nature. While full validation and verification of models and methods, as well as of results, seem infeasible, benchmarking against other similar analyses, plausibility checks and checks against experienced events, if available, may help to support the credibility of the vulnerability assessment and build confidence in the decision making which follows.

The results of simulations typically represented by plots will need to undergo an appropriate process of "zooming" for the necessary deep understanding of the underlying scenarios and the examination of the main influencing parameters/factors. This would be helpful for finding relevant failure combinations and potential artifacts, respectively.

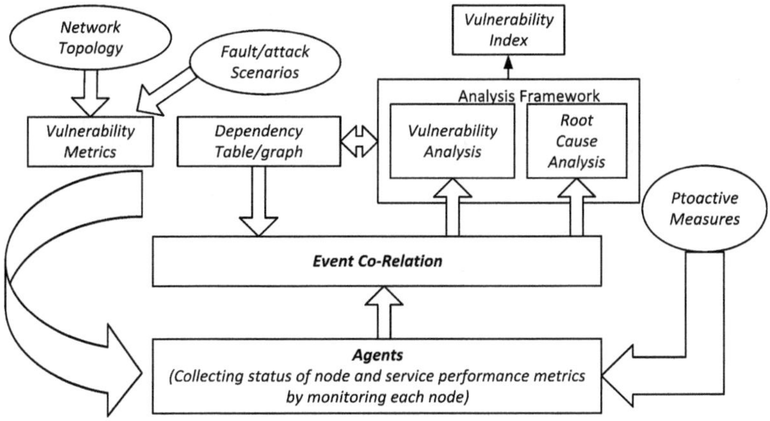

Fig. 5.5 Agent-based vulnerability analysis architecture for the Internet (Qu et al. 2002)

5.2.2.4 Results Assessment and Suggestions for Improvements

The final results of the vulnerability analysis should be assessed regarding their overall credibility. If a distinct need arises to further develop the knowledge base and/or the modeling and simulation techniques, R&D work should be triggered (step 4, Fig. 5.2) and the whole analysis might be delayed.

Depending on the results regarded credible and the related feedbacks from the ordering entity and system owner(s)/operator(s), system improvements (step 5) may be proposed to further reduce and better manage vulnerabilities by all means of provisions. The may include—besides "soft factors"—improved structural and functional design of the system(s), including redundancy and spatial separation, lowered stress level by increased safety margins and modified operational conditions, increased investment into new capacities and/or replacement of old equipment, increased cyber security, adapted emergency, etc.

The final decision about actually implementing the proposals of improvement is left to the "problem owner", possibly after iterating the vulnerability analysis in order to take changes into account and to assess the effectiveness of the proposed improvements and avoid negative feedbacks.

References

Aven T (2010) A unified framework for risk and vulnerability analysis covering both safety and security. Reliab Eng Syst Safe. doi:10.1016/j.ress2006.03.008

Dudenhoeffer DD, Permann MR, Manic M (2006) CIMS: A framework for infrastructure interdependency modeling and analysis. In Proceedings of the 38th conference on winter simulation, Monterey, California, 2006

Eusgeld I, Kröger W (2008) Comparative evaluation of modeling and simulation technique for interdependent critical infrastructures. In: Proceedings of PSAM9, Hong Kong, 18-23 May 2008, pp. 49–57

Eusgeld I, Kröger W, Sansavini G, Schläpfer M, Zio E (2009) The role of network theory and object-oriented modeling within a framework for the vulnerability analysis of critical infrastructures. Reliab Eng Syst Safe 94:954–963. doi:10.1016/j.ress.2008.10.011

Johansson J, Hassel H (2010) An approach for modelling interdependent infrastructures in the context of vulnerability analysis. Reliab Eng Syst Safe. doi:10.1016/j.ress2010.06.010

MIA (2010) Definition of a methodology for the assessment of mutual interdependencies between ICT and electricity generation/transmission infrastructures. Final report, September 2010, Italian National Agency for New Technology, Energy and Environment, Italy

Piwowar J, Châtelet E, Laclémence P (2009) An efficient process to reduce infrastructure vulnerabilities facing malevolence. Reliab Eng Syst Safe 94:1869–1877. doi:10.1016/j.ress2009.06.009

Qu G, Rudraraju J, Modukuri R, Hariri S, Raghavendra CS (2002) A framework for network vulnerability analysis. ITL Lab, the University of Arizona, University of Sothern California

Utne IB, Hokstad P, Vatn J (2009) A structured approach to modeling interdependencies in risk analysis of critical infrastructures. In: Proceedings of the European safety and reliability conference, ESREL 2009, Prague, Czech Republic, 7–10 September 2009

Chapter 6
Methods of Analysis

Challenges to methods for vulnerability analysis have been distillated and approaches, framed into categories, have been explained briefly in the previous chapters. It has also been stated that no all-encompassing method exists but rather an interplay of methods is necessary to provide trustworthy information about vulnerabilities within and among critical infrastructures (CIs). Starting with the evaluation of statistical data this chapter introduces methods in detail which are regarded as most promising to deal with the complex behavior of these systems within screening and, in particular, in depth analysis. The descriptions include conceptual outlines, basic modeling techniques and expected results as well as application cases and assessment of maturity. As human performance plays an important role and must be an integral part of vulnerability analysis suitable methods for human reliability analysis are depicted at the end of this chapter, related to the characteristics of different CIs. Furthermore, a table summarizing capabilities, accountancy of uncertainties, and maturity for practical application is provided in the concluding Chap. 7.

6.1 Evaluation of Statistical Data

Most infrastructures regarded as critically developed have been operated over a long period of time although they experienced major technology and policy changes, i.e., the high-speed train system has little in common with the "old" fast train network and the change in many countries from monopolistic to competitive market structures interfered with operational conditions significantly. Nevertheless useful experience has been gained and, as most systems are continuously operating under self-surveillance, huge datasets are available in principle which can be used for vulnerability investigations. However reliable data are rare because the systematic selection and evaluation of data is not claimed/enforced in many sectors and/or benefit is put in question by many operators.

W. Kröger and E. Zio, *Vulnerable Systems*, DOI: 10.1007/978-0-85729-655-9_6,

Table 6.1 Descriptive statistics for the NERC DAWG data, 1984–2006 (Hines et al. 2008)

	All	≥300 MW	≥50,000 Customers
Total number of events	861	277	320
Mean size (MW)	584	1,706	1,111
Median size (MW)	90	637	274
Standard deviation (MW)	3,272	5,610	5,163
Mean size customers	62,640	288,720	429,180
Median size customers	1,000	71,000	149,750
Standard deviation customers	87,150	1,020,200	1,076,700

Table 6.2 Statistics for data cause categories based on NERC DAWG data (Hines et al. 2008)

	Percentage of events	Mean size (MW)	Mean size (customers)
Earthquake	0.8	1,408	375,900
Tornado	2.8	367	115,439
Hurricane/tropical storm	4.2	1,309	782,695
Ice storm	5.0	1,152	343,448
Lightning	11.3	270	70,944
Wind/rain	14.8	793	185,199
Other cold weather	5.5	542	150,255
Fire	5.2	431	111,244
Intentional attack	1.6	340	24,572
Supply shortage	5.3	341	138,957
Other external cause	4.8	710	246,071
Equipment failure	29.7	379	57,140
Operator error	10.1	489	105,322
Voltage reduction	7.7	153	212,900
Volunteer reduction	5.9	190	134,543

Due to the high degree of criticality of the electric sector information about disturbances is required by authorities in many countries or collected by operators in their own interest. In the US both the Department of Energy (DOE) and the North American Electric Reliability Council (NERC) require that organizations submit reports when sufficiently large disturbances occur within their territories (Form OC-417, DOE 2005). DOE publishes the reports while NERC provides a data base through its Disturbance Analysis working Group (DAWG); Table 6.1 entails descriptive statistics for the time period of 1984–2006. Furthermore an international event base is maintained by the National Institute for the Prevention of Terrorism (www.MIPT.org) focused having on terrorist attacks.

Data bases can be used for various purposes at component or system level, i.e. in the case of electric power systems:

- To provide reliability parameters such as failure and repair rates.
- To identify common components disrupted in the case of outages.
- To categorize initiating events such as natural causes, equipment failure, operator error, demand–supply misbalance, intentional attacks, etc. (see Table 6.2 as an example).

- To identify time series trends and analyze patterns.
- To estimate overall frequency and size of blackout events and find best mathematical fits, e.g. exponential power law distribution.
- To draw general conclusions such as dependence on time-of-year and/or time-of-day.
- To test theoretical approaches and models.
- To provide answers to specific points of interest such as mostly attacked elements, determining factors for the duration of outages.

In order to ensure the reliability of the analyses data may need to be filtered to avoid double counting and to remove irrelevant failures/events depending on the objective of the study. Methods for classical and Bayesian statistics including regression analysis are available for data evaluation (e.g. Zio 2007b). Nevertheless to say that the empirical data must be transferable to the subject of the analysis and of sufficiently high population.

To further illustrate the attractiveness of statistical evaluations the study by (Hines et al. 2008) on trends in the history of a large blackout in the USA may serve as an example. The NERC DAWG data 1984–2006 served as the base inter alia to test the hypothesis that "technology improvements and policy changes have resulted in an observable decrease in the frequency of large blackouts" ($\geq$800 MW) and "that the fit between blackout data and a power law probability distribution is significantly better than the fit to an exponential distribution." Causes were categorized (see Table 6.2), irrelevant data, e.g. related to "voltage and volunteer reduction", were filtered out and finally the cumulative probability distribution of large blackout sizes has been estimated supporting the hypothesis on the superiority of the power-law fit (Fig. 6.1).

On the contrary the hypothesis on decreasing the frequency of major blackout has not been supported, i.e. the blackout frequency has not decreased from 1984 to 2006.

The important result that the frequency of large blackouts is governed by a power law is backed by data from several countries and is interpreted as a clear indication "that the power system being a complex system designed and operated near a critical point" (Dobson et al. 2007).

The above mentioned NERC (domestic) and MIST (international, focused on terrorist attacks) event data bases have also been used to identify key vulnerabilities within the electric power supply system and targets for terrorist attacks. Both data point to the key role of the transmission system, accounting for 60% international attacks (none of them in the USA) and 90% of North American outages (see Table 6.3).

As mentioned before in many sectors and cases sufficient data are not available or not collected and evaluated systematically. To overcome the absence of sound empirically based data proposals have been made to build data banks by making use of "open sources". For example Luiijf et al. (2008) used newspapers and Internet news outlets, if possible augmented by official incident reports, to study CI disruptions and dependencies among them (see Sect. 3.4 for further details).

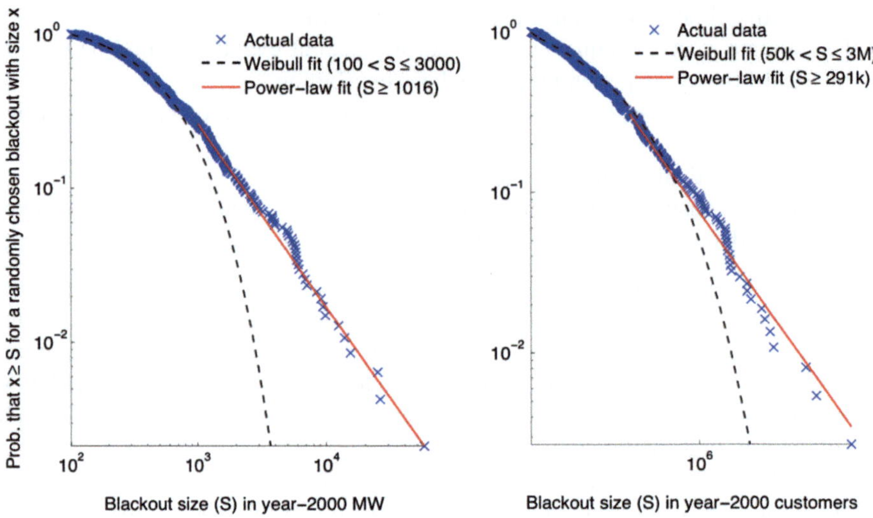

Fig. 6.1 The cumulative probability distribution of blackout sizes in MW (*left*) for events (≥800 MW) and customers (*right*) for events (≥300 k customers). Comparing power-law (*straight lines*) and exponential (*curved lines*) fits to the data show the clear superiority of the power-law fit (Hines et al. 2008)

Table 6.3 Distribution of electric power system components disrupted by type of component for North America (NERC DAWG) and international (MIST) outage databases (Zimmermann et al. 2005)

	North America (*N*)	International (*N*)
Component disrupted		
Transmission lines and towers	182	122
Distribution lines	60	2
Circuit breakers	33	0
Transformers	2,929	77
Substations	21	19
Generation facilities	19	20
Switches and buses	15	0
Other	0	37

Note: For the North American database, more than one component per event could be tabulated in this database so totals do not add to the total number of events in each dataset

Another study (Zimmermann 2004) used an illustrative database of about 100 cases coming from Websites of construction, maintenance or operation accidents, reports of the US National Transportation Safety Board and new media searches "to address the question whether certain combination of infrastructure failures are more common than others". The database includes events from 1990 through 2004 which may also have occurred as a consequence of terrorist attacks and natural hazards, not yet included in the database. The database "only used to illustrate

Table 6.4 Effect rations for selected types of infrastructure (Zimmermann 2004)

Type of infrastructure	No. of times infrastructure (col. 1) *caused* failure of other infrastructure	No. of times infrastructure (col. 1) was *affected by* other infrastructure failures	Ratio of causing versus affected by failure (col. 2 divided by col. 3)
Water mains	34	10	3.4
Roads	25	18	1.4
Gas lines	19	36	0.5
Electric lines	12	14	0.9
Cyber/fiber Optic/telephone	8	15	0.5
Sewers/sewage treatment	8	6	1.3

The results indicate that water mains are more frequent initiators of other infrastructure failures than the reverse (ratio 3.4) whole electric lines are more balanced and have an almost equal chance of disrupting other infrastructure as of being disrupted

how one can conceptualize interdependencies", allows to distinguish between "cause of failure to other infrastructure" and "affected by other infrastructure failures"; the "effect ratio" reflects the extent to which a particular type of infrastructure caused a failure of another type versus being affected by another type infrastructure (see Table 6.4, the letter column 4). The table represents a subset of data reflecting the six types of infrastructure subsectors that accounted for the highest number of failures of other infrastructures.

6.2 Complex Network Theory

In the last decade, a number of studies have focused on the modeling of critical infrastructures (CIs) as complex systems from the standpoint of network theory (Albert and Barabási 2002; Boccaletti et al. 2006). The promise of such research efforts is to unveil insights on network growth mechanisms, points and causes of vulnerability, dynamic behaviors under perturbation, onset of emerging phenomena, etc. The apparent ubiquity of networks leads to a fascinating set of problems common to biological, ecological, technological and social complex systems, concerning how the underlying network topology influences the system behavior and its inherent characteristics of stability and robustness to faults and attacks. The underlying conjecture is that the structure of a system affects its function: for example, the topology of the power grid affects the robustness and stability of power transmission.

From the standpoint of the recent developments in the field of complex systems theory and network analysis, there are two aspects which may allow gaining relevant insights on CIs, if properly analyzed: the study of the topology of the graph representing their structure and the study of their dynamic behavior through functional models reproducing the physical communication processes (mainly flow of some entity, such as electricity, data and vehicles) and the emerging propagation of failures taking place in it.

6.2.1 Conceptual Outline

Recent advances in complex systems theory indicate that many complex systems are hierarchies of networks of interacting components (Strogatz 2001). In this view, the actual structure of the network of interconnections among the components is a critical feature of the system. Indeed, the stability and robustness of these systems depend on the redundant wiring of the functional Web connecting the components; yet, error tolerance and attack robustness are not shared by all redundant networks (Albert et al. 2000).

Unweighted networks, i.e. networks that have a binary nature, where the edges between nodes are either present or not, can be subject to topological analysis (Albert et al. 2000; Strogatz 2001). In a topological analysis, a CI is represented by a graph $G(N, K)$, of N nodes (or vertices) connected by K unweighted (all equal) edges (or arcs). The focus of topological analysis is on the structural properties of the graphs on the global and local scales, e.g. as represented by, respectively, their characteristic path length, L and average clustering coefficient, C (Watts and Strogatz 1998).

Along with a complex topological structure, real networks are equipped with physically heterogeneous nodes and connections, of different capacity and intensity levels (e.g. different impedance and reliability characteristics of overhead lines in electrical transmission networks (Hines and Blumsack 2008; Eusgeld et al. 2009); unequal traffic and accident probabilities on roads (Zio et al. 2008); different routing capacities of the Internet links (Latora and Marchoiri 2005); etc.). In these cases, numerical weights can be assigned to the links and nodes of the representative network, to measure the 'strength' of the connection and node. In this way, the functional behavior of the CI is somewhat embedded into a generalized, but still simple, topological analysis framework.

Furthermore, the interplay between the network structural characteristics and dynamical aspects makes the modeling and analysis very complicated. Functional models have been developed to capture the basic realistic features of CI networks within a weighted topological analysis framework (Motter and Lai 2002; Motter 2004; Zio and Sansavini 2008). These abstract modeling paradigms allow analyzing the system response to cascading failures and can be used to guide a successive detailed simulation focused on the most relevant physical processes and network components. The need for such an analysis tool is even stronger for systems in which the cascade dynamics is rapid and modifications are actuated onto the network in order to mitigate the evolution of the cascade. For example, in electrical power transmission networks a cascade of events leading to a blackout usually occurs on a time scale of minutes to hours and is completed in less than one day (Dobson et al. 2007). The functional analysis of CIs is further complicated by the lack of accurate and complete information. Functional models of CIs require, in fact, the knowledge of a very large amount of data; the network graph must be complemented by a number of information consisting of the technical characteristics of lines and nodes, load requirements, failure probabilities, etc.

These data are often unavailable as they are treated as confidential information from the stakeholders (Rosato et al. 2008).

6.2.2 Modeling Techniques

As mentioned above, the topological analysis for capturing the structural properties of CIs proceeds to model them as graphs whose nodes represent the system units and the links stand for the interactions between directly connected units. In theoretical studies, the connection topology, hereafter also called network structure, of an interconnected system is often assumed to be either a completely regular lattice or a completely random one, to reduce structural related issues, and focus more on the nodes dynamics (Strogatz 2001): Actually, many biological, technological, and social networks lie somewhere between these extremes.

6.2.2.1 Unweighted Networks

The graph $G(N, K)$ representing a CI network is defined by its $N \times N$ adjacency (connection) matrix $[a_{ij}]$ whose entry is:

$$a_{ij} \begin{array}{ll} = 1 & \text{if there is an edge joining vertex } i \text{ to } j \\ = 0 & \text{otherwise} \end{array} \tag{6.1}$$

In practice, the networks are often quite sparse, with $K \ll N \times (N - 1)/2$.

The following constraints define the structural characteristics of different network topologies which may be encountered in the reality of complex systems; the general focus is on sparse, decentralized, connected networks that are neither completely ordered nor completely random (Watts 1999):

(1) The network is numerically large, i.e. made up of a number of interconnected components $N \gg 1$.
(2) The network is sparse, i.e. each component is connected to an average of only $\langle k \rangle \ll N$ other components.
(3) The network is decentralized, i.e. there is no dominant central component to which most others are directly connected; this means that not only the average degree of connection $\langle k \rangle$ must be much less than the network size N but also the maximal degree of connection, $k_{max} \ll N$.
(4) The network is highly clustered.
(5) The network is connected in the sense that any node can be reached by any other node through a finite number of links or edges.

The graph connectivity (or degree) distribution, $P(k)$, can then be evaluated as the probability that a generic node in the network is connected to k other nodes. Existing empirical and theoretical results indicate that complex networks can be divided into two major classes based on their connectivity distribution

$P(k)$ (Boccaletti et al. 2006). The first class of networks is characterized by a connectivity distribution which peaks at the average degree of connection $<k>$ and decays exponentially for large k. The most investigated examples of such exponential networks are the random graph model (Erdos and Rényi 1960) and the 'small-world' model (Watts and Strogatz 1998). Both models represent fairly homogeneous networks in which each node has approximately the same number of links, $k_i \sim <k>$, and in which the degree follows a Poisson distribution, $P(k) = e^{-\langle k \rangle} \times \lambda^k / k!$. Random graphs have been studied deeply within pure mathematics as idealized architectures of gene networks, ecosystems, and the spread of infectious diseases and computer viruses.

The second class comprises inhomogeneous networks, called 'scale-free' (Albert et al. 2000; Crucitti et al. 2003), characterized by highly heterogeneous distributions of a truncated power-law type, $P(k) \sim k^{-\gamma} \varphi(k|\xi)$ where $\varphi(k|\xi)$ introduces a cut-off at some characteristic scale ξ. Three main behaviors can be defined: (a) when ξ is very small, $P(k) \sim \varphi(k|\xi)$, and thus the connectivity distribution is single-scaled, typically corresponding to exponential or Gaussian distributions; (b) as ξ grows, a power law with a sharp cut-off is obtained; (c) for large ξ, networks free of a characteristic scale are obtained. The last two cases have been shown to be widespread in practice and their topological properties have immediate consequences for network robustness and fragility. In these networks, most nodes have very few connections and only a small number of nodes have many connections. It is this inhomogeneous feature that makes a scale-free network error tolerant (i.e. highly robust against random failures like removal of nodes) but is also extremely fragile to attacks like specific removal of the most highly connected nodes. Whereas the probability that a node has a very large number of connections ($k_i \gg <k>$) is practically negligible in exponential networks, highly connected nodes are statistically significant in scale-free networks.

The identification of the network degree distribution, $P(k)$, is a first step in the assessment of the vulnerability characteristics of a CI, providing information on its general response behavior to random failures or targeted attacks. Moreover, it can give insights on the network structure, e.g. a degree distribution peaked at $k = 2$ reveals the mainly sequential structure of the CI under study.

Further characterization of the network structure is sought in terms of single-valued parameters indicating the global and local features of the network, on average. Given the adjacency matrix $[a_{ij}]$ of a network graph $G(N, K)$, it is possible to compute the matrix of the shortest path lengths $[d_{ij}]$ whose entry d_{ij} is the number of edges making up the shortest path linking i and j in the network. The computation can be accomplished in N steps using the Floyd's sequential shortest path iterative algorithm which at each step constructs an intermediate matrix containing the current shortest distance between each pair of nodes, until convergence (Floyd 1962). The fact that G is assumed to be connected implies that the value of d_{ij} is positive and finite $\forall i \neq j$. For studying the global properties of the network topology, the probability distribution $P(d_{ij})$ of the shortest path lengths d_{ij} between any two nodes i and j in the network can be considered. The upper value which d_{ij} can assume is called the diameter of the network and it is used as a measure of the size of the network; it can be thought of as the maximum distance which might be necessary to cover in order to walk from a randomly chosen node to another randomly chosen node (Albert and Barabási 2002). The d_{ij} distribution is useful to get an idea of what in the majority of the cases would be the distance most likely to be covered when moving from one node to another.

The shortest path length distribution is often synthesized by a point value, the 'average' or 'characteristic path length', which represents the average of the shortest distances d_{ij} between all pairs of nodes:

$$L = \frac{1}{N(N-1)} \sum_{i \neq j} d_{ij} \tag{6.2}$$

It represents the average distance which has to be covered to reach the majority of nodes in the graph representing the network system. L is a parameter related to the global structure of the network. The constraint enforcing network connectivity guarantees that L is a truly global statistic. It gives the average minimum distance between any pair of nodes and as such it measures the typical separation between two nodes in the graph (a global property): in a friendship network this would measure the average number of friends in the shortest chain connecting two people (Watts and Strogatz 1998).

The local connectivity of a network can also be synthesized by a single-valued parameter, the so is called average clustering coefficient, C. The clustering coefficient C_i is a local property of node i defined as follows (Albert and Barabási 2002): if node i has k_i neighbors, then at most $k_i \times (k_i - 1)/2$ edges can exist between them; C_i is the fraction of these edges that actually exist; then C is the average of the C_i values:

$$C_i = \frac{\text{Number of edges connecting the neighbours of } i}{\text{Max possible number of edges connecting the neighbours of } i,\ \frac{k_i(k_i-1)}{2}} \tag{6.3}$$

$$C = \frac{1}{N} \sum_i C_i \tag{6.4}$$

Equivalently, C can be regarded as the probability that a pair of vertices u and v are connected given that each one is also connected to a mutual friend w. From the definition, it is clear that C is a measure of the local structure of the network. The largest value that C can attain is 1 for a complete graph (all nodes connected with each other, $<k> = N - 1$) and the smallest is 0, for an empty graph (no connections among the nodes, $<k> = 0$) or a complete sequential graph, a ring, where $k_i = 2 \quad i \in 1, \ldots, N$. Large values of C would be welcome for the robustness of the connectivity: a node removal disconnecting two portions of the system would be overcome by simply passing onto adjacent working nodes through short-range neighboring nodes. C gives the probability that two neighbors of a given node are also neighbors of one another and as such it measures the cliquishness of a typical neighborhood (a local property): in a friendship network,

C_v reflects the probability that friends of v are also friends of each other. In this view, C can be thought of as a simple measure of order: graphs with $C \gg \langle k \rangle / N$ are locally ordered in the sense that nodes with at least one mutually adjacent node are likely to be themselves adjacent (Watts and Strogatz 1998).

In a regular lattice, one has large values of both L and C. In a random graph, both L and C are small. Using the L and C values, it is possible to assess whether the CI structure shares the small-world properties. 'Small-world' networks are characterized by the coincidence of high local clustering of components (much larger than that of an equivalent random graph, $C \gg C_r \sim (<k>/N)$) and short global separation among the clusters (small characteristic path length close to that of an equivalent random graph, $L \sim L_r \sim (\log N / \log <k>)$) (Watts and Strogatz 1998; Watts 1999). As a result, these networks bear two most remarkable properties: local robustness and global accessibility. The robustness comes from dense local clusters (e.g. families, friendship circles, cities and countries in the social context); the global accessibility comes from shortcuts, i.e. edges which connect otherwise separated clusters: it is because of these shortcuts that the world of the complex network seems small and small-world networks exhibit a larger resistance toward targeted attacks, since there is no preferential node in the system.

To test the 'small-worldness' of a network, one has to compare its values of L and C to the corresponding values of a random graph of the same N and K. A random graph of N nodes is built by considering that each possible edge between two given nodes is present with some probability p; then, the average number of edges (the average degree of connection) of a given node in the graph is $\langle k \rangle \sim Np$ and the connectivity distribution $P(k)$ follows a Poisson distribution which peaks at $\langle k \rangle$. Thus, this so called Erdos–Renyi (ER) random graph (Erdos and Rényi 1960) is fairly well characterized by the parameter $\langle k \rangle$ and displays a phase transition at a critical average degree $k_c = 1$: at this critical value, a giant component forms; for $\langle k \rangle > k_c$, a large fraction of the nodes are connected in the network whereas for $\langle k \rangle < k_c$ the system is fragmented in small sub-Webs. The importance of this phenomenon is obvious in terms of the collective properties that arise at the critical point: communication among the whole system becomes available. Besides, the transition occurs suddenly, implying 'innovation', and takes place at a low cost in terms of the number of required links: since the only requirement in order to reach full communication is to have a given (small) number of links per node, once the threshold is reached, order can emerge "for free".

However, there are two limitations which somewhat limit the practical application of the topological indicators L and C for characterizing real network systems (Latora and Marchiori 2001):

(1) They are ill-defined if:

- The network is not fully connected, i.e. some nodes are not connected to the remaining part of the network ($L = \infty$).
- Some nodes have only one neighbor, i.e. $k_i = 0$ ($C_i = 0/0$).

(2) They retain only the topological information on the existence or absence of a link, with no reference to the physical length and capacity of the link. In other words, they are applicable only to unweighted networks.

Regarding the role that an element plays in a network, various measures of the importance of a network node, i.e. of the relevance of its location in the network with respect to a given network performance, have been introduced. In social networks, for example, the so-called centrality measures are introduced as importance measures to qualify the role played by an element in the complex interaction and communication occurring in the network. The term 'importance' is then intended to qualify the role that the presence and location of the element plays with respect to the average global and local properties of the whole network. Classical topological centrality measures are the degree centrality (Nieminen 1974; Freeman 1979), the closeness centrality (Freeman 1979; Sabidussi 1966; Wasserman and Faust 1994), the betweenness centrality (Freeman 1979), and the information centrality (Latora and Marchiori 2007). They specifically rely only on topological information to qualify the importance of a network element.

The topological degree centrality, C^{D}, gives the highest score of importance to the node with the largest number of first neighbors. This agrees with the intuitive way of estimating the influence of a node in a graph from the size of its immediate environment. Quantitatively, the topological degree centrality is defined as the degree of a node, normalized over the maximum number of neighbors this node could have: thus, in a network of N nodes, the topological degree centrality of node i, C_i^{D}, is defined as:

$$C_i^{\mathrm{D}} = \frac{k_i}{N-1} = \frac{\sum_{j \in G} a_{ij}}{N-1} \quad 0 \leq C_i^{\mathrm{D}} \leq 1 \tag{6.5}$$

where k_{i} is the degree of node i and $N-1$ is the normalization factor introduced to account for the fact that a given node i can at most be adjacent to $N-1$ other nodes. The running time required for computing C^{D} for all nodes is $O(N)$.

The topological closeness centrality, C^{C}, captures the idea of speed of communication between nodes in such a way that the node which is "closest" to all others receives the highest score. In other words, this measure allows identifying the nodes which on average need fewer steps to communicate with the other nodes, not only with the first neighbors. Because this measure is defined as "closeness", quantitatively the inverse of the node's mean distance from all the others is used. If d_{ij} is the topological shortest path length between nodes i and j, i.e. the minimum number of edges traversed to get from i to j, the topological closeness centrality of node i is:

$$C_i^{\mathrm{C}} = \frac{N-1}{\sum_{j \in G} d_{ij}} \quad 0 \leq C_i^{\mathrm{C}} \leq 1 \tag{6.6}$$

Note that also this measure is normalized to assume values in the interval [0,1].The running time required for computing C^{C} for all the nodes by means of the Floyd algorithm is $O(N^3)$.

The topological betweenness centrality, C^{B}, is based on the idea that a node is central if it lies between many other nodes, in the sense that it is traversed by many of the shortest paths connecting pairs of nodes. The topological betweenness centrality of a given node i is quantitatively defined as:

$$C_i^{\mathrm{B}} = \frac{1}{(N-1)(N-2)} \sum_{j,k \in G, j \neq k \neq i} \frac{n_{jk}(i)}{n_{jk}} \quad 0 \leq C_i^{\mathrm{B}} \leq 1 \tag{6.7}$$

where n_{jk} is the number of topological shortest paths between nodes j and k, and $n_{jk}(i)$ is the number of topological shortest paths between nodes j and k which contains node i. Similarly to the other topological centrality measures, C_i^{B} assumes values between 0 and 1 and reaches its maximum when node i falls on all geodesics (paths of minimal length between two nodes). The running time required for computing C^{B} for all nodes by means of the Floyd algorithm is $O(N^3)$.

6.2.2.2 Weighted Networks

In practice, the model of a realistic network could be weighed (e.g. by its physical characteristics of reliability and capacity), non-sparse, and non-connected. Thus, to account also for the physical properties of the systems, network efficiency measures have been introduced as complements to the classical topological indicators such as the characteristic path length L and the clustering coefficient C (Latora and Marchiori 2001). In addition to the adjacency matrix $[a_{ij}]$, defined as for the unweighted graph, an additional matrix $[l_{ij}]$ of weights, e.g., physical distances (Latora and Marchiori 2001), failure/accident probabilities (Zio et al. 2008; Eusgeld et al. 2009), and 'electrical' distances (Hines and Blumsack 2008), can be introduced to describe the network. Of course, in the case of an unweighted network, $l_{ij} = 1 \; \forall i \neq j$.

On the basis of both $[a_{ij}]$ and $[l_{ij}]$, the matrix of the shortest path lengths $[d_{ij}]$ is computed: the length d_{ij} of the shortest path linking i and j in the network is the smallest sum of the physical distances throughout all the possible paths from i to j.

Assuming that the network system is parallel, i.e. that every node concurrently sends information through its edges, a measure of efficiency in the communication between nodes i and j can be defined, inversely proportional to the shortest distance (Latora and Marchiori 2001). Thus, the network is characterized also by an efficiency matrix $[\varepsilon_{ij}]$, whose entry is the efficiency in the communication between nodes i and j:

$$\begin{aligned} \varepsilon_{ij} &= \frac{1}{d_{ij}} \quad \text{if there is at least one path connecting } i \text{ and } j \\ &= 0 \quad \text{otherwise } (d_{ij} = \infty) \end{aligned} \tag{6.8}$$

The average efficiency of $G(N, K)$ is then

$$E_{\mathrm{glob}}(G) = \frac{\sum_{i \neq j \in G} \varepsilon_{ij}}{N(N-1)} = \frac{\sum_{i \neq j \in G} \frac{1}{d_{ij}}}{N(N-1)} \tag{6.9}$$

This quantity plays the same role of L in defining the network connection characteristics on a global scale. The fundamental difference is that E_{glob} is the efficiency of a parallel network of nodes which concurrently exchanges packets of information, whereas $1/L$ measures the efficiency in a *sequential* system where only one packet of information at the time goes along the network. Thus, $1/L$ represents well the efficiency of unweighted networks where no difference is made on the distances in the graph.

For comparison of different network systems, it is useful to normalize $E_{glob}(G)$ by considering the ideal, fully connected network $G_{id}(N)$ in which all N nodes of the network are connected and which, thus, contains $N(N-1)/2$ edges. Such a system propagates the information in the most efficient way since $[d_{ij}] = [l_{ij}]\ \forall i \neq j$. The corresponding (maximum) value of global efficiency is:

$$E_{glob}(G_{id}) = \frac{\sum_{i \neq j \in G_{id}} \frac{1}{l_{ij}}}{N(N-1)} \tag{6.10}$$

By dividing Eq. 6.9 by 6.10, one obtains a normalized value of global efficiency for the graph $G(N, K)$, which for simplicity of notation is still denoted as $E_{glob}(G)$ and is such that $0 \leq E_{glob}(G) \leq 1$.

One can also quantify the local properties of the graph $G(N, K)$ by specializing the definition of the average efficiency (Eq. 6.9) on the subgraph G_i of the neighbours of each node i in the network:

$$E(G_i) = \frac{\sum_{n \neq m \in G_i} \varepsilon_{nm}}{k_i(k_i - 1)} \tag{6.11}$$

Averaging the efficiency of the local neighborhoods of all nodes in the network one can define a measure of the network local efficiency:

$$E_{loc}(G) = \frac{1}{N} \sum_{i=1 \in G}^{N} E(G_i) \tag{6.12}$$

Since $i \notin G_i$, this parameter reveals how much the system is fault tolerant in that it shows how efficient the communication between the first neighbors of i remains when i is removed.

The local efficiency $E_{loc}(G)$ plays a role similar to the clustering coefficient C in measuring how well connected is a network. It can be shown that when most of the local subgraphs G_i of a graph G are not sparse, C gives a good approximation of $E_{loc}(G)$ (Latora and Marchiori 2001).

The definition of the small world behavior can then be rephrased in terms of the information flow: small world networks are characterized by high values of both E_{glob} and E_{loc}, i.e. high efficiency in both global and local communications.

Average global and local topological efficiency measures form the output of this analysis. It is worth mentioning that for $[a_{ij}] = [l_{ij}]$ the weighted analysis coincides with the unweighted topological analysis in that all edges are equally weighted ones.

Various measures that allow for a statistical characterization of weighted networks can be developed, encompassing a range of possible applications. For example, the local and global reliability characteristics of a complex network system have been considered within a framework of vulnerability assessment of CIs (Zio 2007a, b; Zio and Sansavini 2007). By considering the 'reliability distances' among network nodes in terms of the probabilities of failure of the interconnecting links, global and local reliability efficiency indicators can be defined for use in the analysis of the robustness and vulnerability of network systems and thus for their optimal design, operation and management. Two different definitions are introduced, depending on how the 'length' d_{ij} of the shortest path linking two generic nodes i and j in the network is defined.

Let q_{ij} be the probability that edge ij is incapable of transmitting information between nodes i and j and $p_{ij} = 1 - q_{ij}$ be the probability of successful transmission along the edge. On the contrary, nodes are considered infallible for simplicity of illustration. It is customary to call failure probability of edge ij the former and reliability of edge ij the latter. In addition to the adjacency matrix $[a_{ij}]$, defined the same as for the unweighted graph, an additional matrix $[q_{ij}]$ (or the complementary $[p_{ij}]$) is introduced to describe the network.

The failure probability $Q_{\gamma_{ij}}$ and the reliability $P_{\gamma_{ij}}$ of a generic path γ_{ij} of independent edges connecting nodes i and j are simply computed as

$$\begin{aligned} Q_{\gamma_{ij}} &= 1 - \prod_{mn \in \gamma_{ij}} p_{mn} = 1 - \prod_{mn \in \gamma_{ij}} (1 - q_{mn}) \\ P_{\gamma_{ij}} &= \prod_{mn \in \gamma_{ij}} p_{mn} = \prod_{mn \in \gamma_{ij}} (1 - q_{mn}) \end{aligned} \tag{6.13}$$

On the basis of both $[a_{ij}]$ and $[q_{ij}]$ (or the complementary $[p_{ij}]$), the matrix of the shortest (most reliable) path lengths $[d_{ij}]$ is computed: the 'length' d_{ij} of the shortest (most reliable) path linking i and j in the network is defined as:

$$d_{ij} = \min_{\gamma_{ij}} \left(\ln Q_{\gamma_{ij}} \right) = \min_{\gamma_{ij}} \left(- \sum_{mn \in \gamma_{ij}} \ln p_{mn} \right) = \min_{\gamma_{ij}} \left(- \sum_{mn \in \gamma_{ij}} \ln(1 - q_{mn}) \right) \tag{6.14}$$

where the minimization is done with respect to all paths γ_{ij} linking nodes i and j and the sum extends to all the edges of each of these paths. Note that $0 \leq d_{ij} \leq \infty$, the lower value corresponding to the existence of a perfectly reliable path connecting i and j (all edges along such path cannot fail, i.e., $p_{mn} = 1, q_{mn} = 0 \; \forall mn \in ij$) and the upper value corresponding to the situation of no paths connecting i and j (i.e. in all paths from i to j there is at least one that certainly fails $p_{mn} = 0, q_{mn} = 1$).

In this view, the efficiency of communication ε_{ij} between nodes i and j can be defined, analogously to before, as being inversely proportional to the shortest distance Eq. 6.14. Note that coherently, the efficiency $\varepsilon_{ij} = 0$ when there is no path in the graph linking nodes i and j $\left(Q_{\gamma_{ij}} = 1, P_{\gamma_{ij}} = 0 \; \forall \gamma_{ij}, \; d_{ij} = \infty \right)$. The efficiency matrix ε_{ij} can then be introduced as before Eq. 6.8 and the average global and local efficiencies computed (Eqs. 6.9 and 6.12).

As for the normalization of the global efficiency, this can be again done with respect to the ideal, fully connected network G_{id} in which all N nodes of the network are connected by $N(N-1)/2$ edges. A value of failure probability q_{ij} (or alternatively of reliability p_{ij}) must be assigned to each edge $ij \in G_{\mathrm{id}}$. In this respect, an obvious choice would be to consider perfect, non-failing edges ($p_{mn}=1$, $q_{mn}=0\ \forall mn$). However, this would lead to null distances and infinite efficiencies at each connection ($d_{ij}=0$, $\varepsilon_{ij}=\infty\ \forall ij \in G_{\mathrm{id}}$). Then, the maximum global efficiency $E_{\mathrm{glob}}(G_{\mathrm{id}})=\infty$ and the normalized global efficiency of any network G would be zero. To solve this problem, a possibility is to assign to all edges of the fully connected network a very small (large), 'ideal' value of failure probability (reliability) close to zero (unity), say $10^{-\alpha}$ $(1-10^{-\alpha})$, with $\alpha > 1$, to be interpreted as the optimal value technologically achievable for the edge components in the kind of network under analysis.

The previous definitions of global and local network reliability efficiencies arise naturally from the reliabilities of the paths connecting the network nodes. Yet, the logarithmic form of these indicators makes them little sensitive to changes in both the network topology and in the reliability values of its elements (Zio 2007a).

An alternative, more sensitive definition of the 'length' d_{ij} of the shortest (most reliable) path linking i and j in the network is as follows:

$$d_{ij} = \min_{\gamma_{ij}}\left(\frac{1}{\prod_{mn\in\gamma_{ij}} p_{mn}}\right) = \min_{\gamma_{ij}}\left(\frac{1}{\prod_{mn\in\gamma_{ij}}(1-q_{mn})}\right) \tag{6.15}$$

where the minimization is done with respect to all paths γ_{ij} linking nodes i and j and the product extends to all the edges of each of these paths. Note that $1 \le d_{ij} \le \infty$, the lower value corresponding to the existence of a perfectly reliable path connecting i and j (all edges along such path cannot fail, i.e. $p_{mn}=1, q_{mn}=0\ \forall mn \in ij$) and the upper value corresponding to the situation of no paths connecting i and j (i.e. in all paths from i to j there is at least one that certainly fails $p_{mn}=0, q_{mn}=1$).

Correspondingly, the efficiency of communication ε_{ij} between nodes i and j, given as before by the inverse of the shortest distance Eq. 6.15, is such that $0 \le \varepsilon ij \le 1$, the lowest value of efficiency corresponding to the situation when there is no path in the graph linking nodes i and j $\left(Q_{\gamma_{ij}}=1,\ P_{\gamma_{ij}}=0\ \forall\gamma_{ij},\ d_{ij}=\infty\right)$ whereas the highest value is attained when there is at least one perfect path γ_{ij} in the graph, which connects nodes i and j through a sequence of non-failing edges $(Q_{\gamma_{ij}}=0,\ P_{\gamma_{ij}}=1)$. The efficiency matrix ε_{ij} can then be introduced as before (Eq. 6.8) and the average global and local efficiencies computed (Eqs. 6.9 and 6.12).

As mentioned before, an important issue for the protection of CIs is that of determining the critical elements in the network. Methods have been proposed to evaluate the importance of an edge of the network by considering the drop in the network's performance, $E_{\mathrm{glob}}(G)$, caused by its deactivation (Latora and Marchiori 2005). In practice the redundancy of an element is checked by calculating the performance of the perturbed network, G^*, and comparing it with the original one, G. Notice that the element can be either a single node or edge, or a group of nodes/edges in case multiple attacks are simulated. The output of the analysis is a ranking of the elements of the network that should be of primary concern for any policy of protection from random failures or terrorist attacks. Furthermore, the importance of an improvement could be measured by the increase in the network's performance caused by it. When some edges are removed from the network, the shortest paths between nodes change due to forced detours around the blockages. In this view, the vulnerability of the network is defined in terms of the degradation in the global efficiency of the network due to the disconnection of a set of its links:

$$V^* = \frac{E_{\text{glob}}(G) - E_{\text{glob}}(G^*)}{E_{\text{glob}}(G)} \tag{6.16}$$

By construction, V^* takes values in the range [0, 1].

With respect to the role played by individual elements in the network, a topological information centrality, C^{I}, can be introduced to relate a node importance to the ability of the network to respond to the deactivation of the node. In this view, the network performance is measured by the network topological efficiency $E[G]$ defined as Eq. 6.9.

The topological information centrality of node i is defined as the relative drop in the network topological efficiency caused by the removal of the edges incident in i:

$$C_i^I = \frac{\Delta E(i)}{E} = \frac{E[G] - E[G'(i)]}{E[G]} \quad 0 \leq C_i^I \leq 1 \tag{6.17}$$

where $G'(i)$ is the graph with N nodes and $K - k_i$ edges obtained by removing from the original graph G the edges incident in node i. An advantage of using the efficiency to measure the performance of a graph is that $E[G]$, contrary to L, is finite even for disconnected graphs, also C^I is normalized in the interval [0,1], by definition. The running time required for computing C^I for all nodes by means of the Floyd algorithm is $O(N^4)$.

More generally, for weighted networks it is possible to extend the previously defined centrality measures, so as to account for the physical characteristics of the network arcs. For example, for the case of reliability-weighted networks, the reliability degree centrality, RC^{D}, of node i in a network of N nodes is defined as:

$$\text{RC}_i^{\text{D}} = \frac{k_i \sum_{j \in G} p_{ij}}{(N-1)^2} \quad 0 \leq \text{RC}_i^{\text{D}} \leq 1 \tag{6.18}$$

where k_i is the degree of node i and p_{ij} is the reliability of edge ij. Differently from Eq. 6.5, the normalization factor $(N - 1)^2$ is introduced here to account for the fact that $\max(k_i) = N - 1$ when the node i is fully connected and $\max\left(\sum_{j \in G} p_{ij}\right) = N - 1$ when all the $N - 1$ edges are fully reliable ($p_{ij} = 1, \forall j \in G$). Thus, the measure RC^{D} is normalized in the interval [0,1].

The reliability closeness centrality, RC^C, measures to which extent a node i is near to all other nodes along the most reliable paths and is defined in the same way as its topological analog C_i^C Eq. 6.6. Also RC^C assumes values in the interval [0,1].

The reliability betweenness centrality, RC^B, is based on the idea that a node is central if it lies between many other nodes, in the sense that it is traversed by many of the most reliable paths connecting pairs of nodes; it is defined in the same way as its topological analog C_i^B Eq. 6.7, in which n_{jk} is replaced by rn_{jk} (number of most reliable paths between nodes j and k) and $n_{jk}(i)$ is replaced by $rn_{jk}(i)$ (number of most reliable paths between nodes j and k that contain node i). Also this measure is normalized in the range [0,1].

For the reliability information centrality, RC^I, the network performance is measured by the reliability efficiency $E[G]$ of the graph G defined as in Eq. 6.9. The reliability information centrality of node i, RC_i^I, is defined as its topological analog C_i^I (Eq. 6.17), but with the network reliability efficiency replacing the topological efficiency. RC^I is also normalized in the interval [0,1].

The running times required for computing the above reliability centrality measures, RC^D, RC^C, $RC^B$and RC_i^I are the same as those for the topological cases.

6.2.3 Failure Cascades Modeling

Up to this point, the analysis has dealt with static measures evaluating the structural properties of a CI and the identification of the central components or the effects of the removal of an edge in a network have been analyzed without considering the time evolution or space propagation of network failures. On the contrary, a thorough vulnerability analysis of a CI must take into account the spreading processes of failure cascades, which, despite being triggered by an initiating local disturbance (e.g. a single failure of a component), can affect the whole network system and, possibly, disrupt the service it provides.

Since network topology can have a strong influence on the failure spreading mechanism, the analysis of the cascade failure evolution must deal with the mutual interplay between the system's dynamics and structural complexity.

Various abstract models of cascading failures have been applied to simulate the *propagation process in single CIs*, differing for both the logic of redistribution of the failure load and the nature of the cascade triggering event, i.e. either a random failure or a targeted intentional attack (Motter and Lai 2002; Dobson et al. 2005; Zio and Sansavini 2008). The choice of the most suitable algorithm for modeling the spreading process taking place in a given CI must be performed carefully, considering the type of service provided by the CI, as further explained below.

Some examples are given in the following, considering failure-free edges to focus on the failure propagation dynamics over the network nodes (schemes of edge removal can be straightforwardly implemented). In the abstract failure cascade modeling presented (Zio and Sansavini 2008), the spreading variable of interest is typically normalized in the range [0, 1], for generality purposes; for application to specific CIs, the normalization is dropped and the physical nature of the service provided by the CI is considered to tailor the abstract model to the real system.

In the study of the failure cascade propagation mechanism triggered by random failures or targeted attacks, the CI is still modeled as a graph $G(N, K)$, like in the static analysis. The K connections establish the communication pattern on which the quantities of relevance are transferred in the network and affect the spreading dynamics, but they are assumed not subject to failures during the cascade propagation.

In many models, loads are assigned to nodes irrespective of the connectivity pattern of the system which only affected the direction of load propagation following a component failure. These models better reflect the behavior of systems like the power distribution ones, where the load at each substation does not follow directly from the number of overhead lines pointing to it. On the other hand, in systems like information networks the load on each component, e.g. a router or a hypernode, can be thought as dependent on the number of links transiting through it, i.e. its betweenness centrality (Latora and Marchiori 2005).

To model the cascade dynamics arising from a random failure or a targeted intentional attack at network components, let us assume that at each time step one unit of the relevant quantity processed by the network, which can be for example information, is exchanged along the shortest paths connecting every pair of components. The load at a component is then the total number of shortest paths passing through that component (Batagelj 1994; Newman and Girvan 2004), i.e. its betweenness centrality evaluated in Eq. 6.7, $L_j = C_j^{\mathrm{B}}$, to be compared with its capacity which is the maximum load that the component can handle. In man-made networks, the capacity is usually limited by technological limitations and economic considerations. Thus, it seems reasonable to assume that the capacity C_j of the component j is dimensioned proportionally to its nominal load L_j at which it is designed to operate initially (Motter and Lai 2002),

$$C_j = \alpha \times L_j \quad j = 1, 2, \ldots, N \tag{6.19}$$

where the constant $\alpha > 0$ is the capacity tolerance parameter, assumed equal for all components. When all the components are working, the network operates without problems in so far as $\alpha > 0$. On the contrary, the occurrence of component failures leads to a redistribution of the shortest paths in the network and, consequently, to a change in the loads of the surviving components. If the load on a component increases beyond capacity, the component fails and a new redistribution of the shortest paths and loads follows, which, as a result, can lead to a cascading effect of subsequent failures.

The importance of the cascade effect with respect to intentional attacks stems from the fact that a large damage can be caused by the attack on a highly central single component. Obviously, in general more links render a network more resistant against cascading failures, but this increases the cost of the network (Cadini et al. 2009).

When looking at the potential for cascading processes triggered by the removal of a single component, two situations are expected: if prior to its removal the component is operating at a relatively small load (i.e. if a small number of shortest paths go through it), its removal will not cause major changes in the balance of loads and subsequent overload failures are unlikely; however, when the load of the component is relatively large, its removal is likely to affect significantly the loads of other components and possibly start a sequence of overload failures. Intuitively, then, the following behavior is expected (Motter and Lai 2002): global cascades occur if the network exhibits a highly heterogeneous distribution of loads and the removed component is among those with highest loads; otherwise, cascades are not expected.

In the modeling scheme adopted, the distribution of loads is in turn highly correlated with the distribution of links: networks with heterogeneous distribution of links are expected to be heterogeneous with respect to the load, so that on average components with a large number of links will have high loads. This behavior confirms the robust-yet-fragile property of heterogeneous networks, which was first observed in (Albert et al. 2000) with respect to the attack on several components.

To quantitatively show the above described behavior, a component among those with largest degrees in the network is selectively attacked. Such a component is removed together with all its links and the redistribution of shortest paths and loads is performed. The damage caused by the resulting cascade is quantified in terms of the number of network components which remain in operation.

In order to delve into the failure cascade unfolding, the four cascade indicators, Eqs. 6.21–6.24 introduced in the following Sect. 6.4.3.1, can be used. In particular, in an effort to enrich the effectiveness of the static analysis of CIs and to bridge the findings of the static and dynamic analyses, the rankings obtained with the different cascade criticality indicators can be compared with classical topological centrality measures (Cadini et al. 2009).

The degree and betweenness centralities, Eqs. 6.5 and 6.7, which account for the number of connections pointing to a component and for the number of shortest paths passing through a component, respectively, appear to play a major role in identifying those network components which most contribute to the failure propagation process. The betweenness centrality measure, only partially highlights those components which most contribute in determining large-sized failure cascades.

Resorting solely to fast topologic centrality measure evaluations, it seems then possible to preliminary rank the criticality of single nodes with respect to the failure propagation phenomena which can occur in the CI, without the need to run time-expensive simulations with detailed codes.

6.2.4 Expected Results

The two main outputs of a CI vulnerability assessment are the quantification of system vulnerability indicators and the identification of critical elements. The information they provide is complementary: while vulnerability indicators are parameters encompassing the static or dynamic characteristics of the whole system, the identification of critical elements provides rankings of components criticalities with respect to their connectivity efficiency or their contributions to the propagation of failure through the network.

The determination of the global and local indicators of the topological and weighted characteristics give valuable information on the connectedness efficiency of the network system of a CI and can be useful when compared with reference, ideal values.

The pure topological structure can be complemented with weights extrapolated from the physical features of the CI. As an example, reliability and electrical 'distances' can be combined in the vulnerability assessment of an electrical transmission system and used as weights for the network arcs of the graph representative of the system (Zio and Golea 2010); by so doing, the vulnerability and centrality measures evaluated in the weighted analysis encompass the information on the physics of the service provided by the CI under analysis. Further, the analysis of the vulnerability of the network in terms of the degradation of its global efficiency due to the disconnection of a set of links as in Eq. 6.16, which is the homologous of the information centrality for nodes of Eq. 6.17, allows ranking the network elements (arcs or nodes) with respect to their role in the network global communication efficiency.

From the modeling of a cascading failure propagation process, a useful output is the average loss of service in terms of C_i^{B} versus the tolerance parameter, α, from which the minimum flow-carrying capacity of the network, α, ensuring a minimum connectivity level i after the cascades spreads in the network can be identified. This minimum capacity α identifies the safe and critical failure-prone working conditions for a network, since the disruption following a cascading failure affects the system in a catastrophic way if the system in operating under this limit. Therefore, α can be used as a system vulnerability indicator for CIs.

Furthermore, following the modeled cascade spreading process step by step, indicators can be evaluated for each component, such as the frequency of its participation to a cascade, the average time before its entrance into a cascade, the average duration and final size of the cascade emerging from its failure, which identify the criticality of the components with respect to their contribution to the development of cascading failures. To the same aim, classical measures of topological centrality can be used for components criticality ranking with respect to their contribution to the failure cascade process in network systems.

6.2.5 Exemplary Applications

The methods of vulnerability analysis of CIs via complex network theory are shown with respect to their application to electrical infrastructures (Eusgeld et al. 2009; Zio and Sansavini 2011a, b). The topological and the weighted analyses are applied to the Swiss 220 kV/380 kV high voltage transmission system, while the

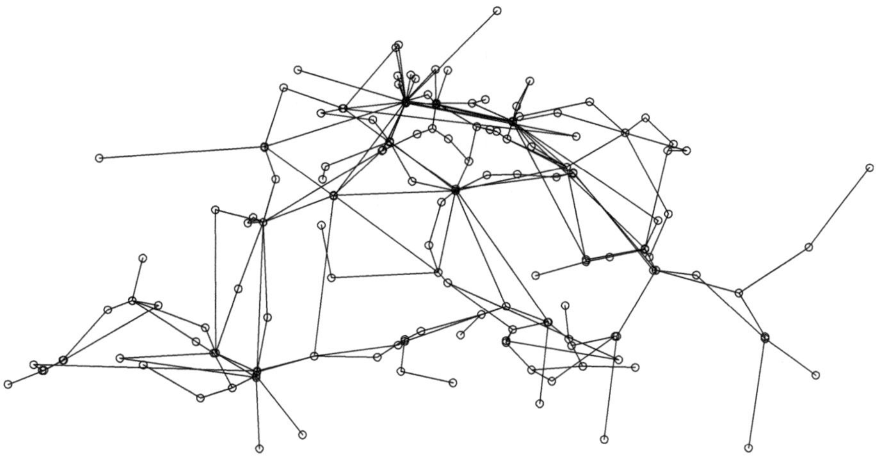

Fig. 6.2 The 220 kV/380 kV Swiss power transmission network

dynamic analysis of failure cascade propagation is applied to the topological network of the 380 kV Italian power transmission network (TERNA 2002; Rosato et al. 2007) aiming at ranking the criticality of its components in the cascading failure propagation process, and to a modified literature case study (Grigg et al. 1996), with the aim of identifying the interdependency features and defining related cascade-safe operational margins.

6.2.5.1 Example of Topological and Weighted Analysis

The reference system for the static topological analysis and for the weighted analysis (Eusgeld et al. 2009) is the Swiss 220 kV/380 kV high voltage transmission system (Fig. 6.2) which consists of a single control area.

The system is made up of $N = 161$ nodes (busbars) connected by $K = 219$ overhead lines (double lines are modeled by single lines). The system is modeled as a stochastic, undirected, connected graph $G(N, K)$ in which each substation is transposed into a node, linked by edges representing the overhead lines connecting two subsequent substations (Fig. 6.2).

Depending on the analysis, each edge is either unweighted (weight $= 1$), for a purely topological analysis, or weighted by its reliability for an analysis of the power transmission efficiency.

The stochastic failure behavior of the power transmission system is described by introducing for the generic overhead line connecting substation i to substation j, a constant annual failure rate λ_{ij} per km. Such failure rate represents the number of failures occurred in 1 year along 1 km of overhead line connecting a given pair of substations i and j. For the sake of simplicity, it is assumed that $\lambda_{ij} = \lambda$ for the whole power transmission network. Given the assumption of constant failure rate, the annual reliability of the line connecting nodes i and j is

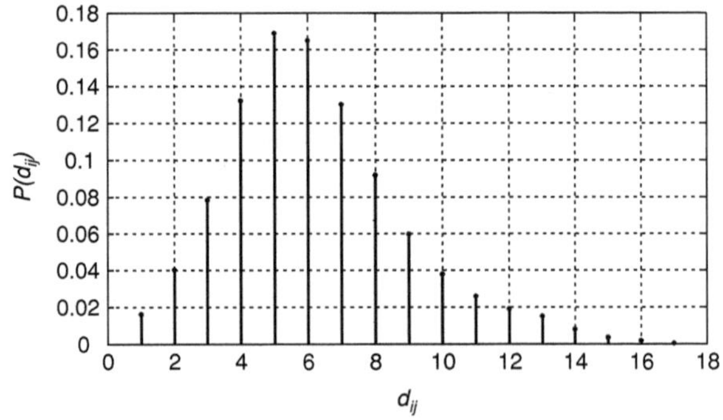

Fig. 6.3 Shortest path length distribution for the Swiss power transmission network

$$p_{ij} = e^{-\lambda \cdot l_{ij}} \tag{6.20}$$

where l_{ij} is the length of the line.

We shall refer to this quantity as the reliability of edge *ij* and call failure probability of edge *ij* its complement to one, $q_{ij} = 1 - p_{ij}$. Thus, in addition to the adjacency matrix $[a_{ij}]$, the additional matrix $[p_{ij}]$ (or the complementary $[q_{ij}]$) is introduced to describe the failure behavior of the power transmission system.

Some cautious words should be spent on the crude homogeneity assumption that $\lambda_{ij} = \lambda$ for all lines in the power transmission network. A single value of annual failure rate per 100 km of overhead line was available from VSE-AES Statistik (2005). This value takes into account atmospheric effects, external effects (due to external human actions) and operational margins (unexpected overloads, wrong operations, planned maintenance, failure of the material). With the homogeneity assumption of all failure rates being equal, the probabilities of failure of the various overhead lines differ only due to their length. In reality, other factors should be taken into account in the reliability evaluation, like the geographical position or the age of the material of the line and its usage.

Both topological and reliability efficiency analyses have been carried out for assessing the power transmission network properties and transmission performances.

Figure 6.3 shows that the shortest path length distribution has a tail up to $d_{ij} = 17$, implying that one has to pass at most through 17 nodes for the power to be transmitted from one point to another in the network. This value is the diameter of the network. The largest portion of the distribution is concentrated around values of $d_{ij} = [3,\ 8]$ and the distribution peaks at $d_{ij} = 5$, implying that the connectivity of this network is high. A characteristic path length $L = 6$ is found. This clearly reflects the d_{ij} distribution and confirms that the network has good global connectivity properties.

The degree distribution, plotted in Fig. 6.4, peaks at about $k = 2$ but has quite large values also for $k > 2$. This implies that a failed substation disconnected from the network can easily be overtaken through other paths in the system. Nodes with $k = 1$ are the boundary substations of the Swiss power transmission network.

A direct measure of the clustering coefficient of the power transmission network gives the rather small value of $C = 7.79 \times 10^{-2}$. The predominant series structure

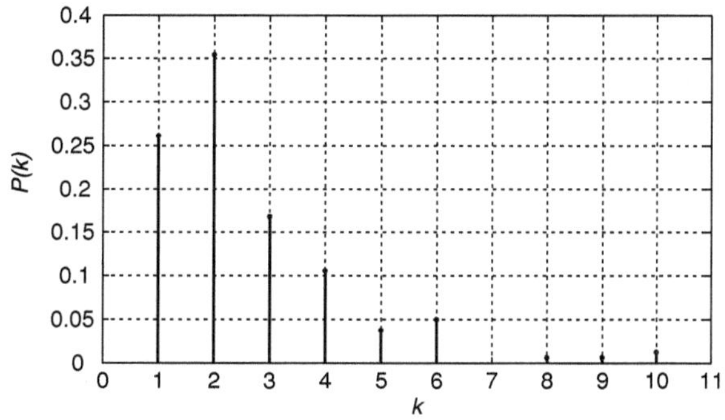

Fig. 6.4 Degree distribution for the Swiss power transmission network

Table 6.5 Topological and reliability efficiencies $E_{glob}(G)$, $E_{loc}(G)$

	Topological efficiency	Reliability efficiency
Global efficiency, $E_{glob}(G)$	20.5×10^{-2}	9.30×10^{-2}
Local efficiency, $E_{loc}(G)$	7.89×10^{-2}	4.72×10^{-2}

of the network is responsible for the large number of sparse subgraphs around the nodes, a phenomenon which leads to the small values of the average clustering coefficient.

One could ask whether there is any reason for a power transmission network to have high clustering. In principle, this would be welcome for the robustness of the service: a fault stopping a portion of a transmission line would be overcome by simply passing onto other working points of the system. Actually, this rerouting of the power flux can be accomplished simply by nodes with $k = 3$ which do not imply large system clustering.

In Table 6.5, the values of global and local reliability efficiencies are shown. They are compared with the topological efficiencies which are the upper values of reliability efficiencies for perfectly reliable overhead lines. The topological efficiencies account only for the topological connectivity pattern in the network (unweighted links).

In particular, the global topological efficiency is equal to one for a fully connected graph in which every node is directly connected with any other node. In Fig. 6.5, the distribution $P(E(G_i))$ of the average efficiencies on the subgraph G_i Eq. 6.11 is presented. Note that almost all subgraphs are sparse, thus giving null contribution to the average local efficiency. This distribution reflects the previously discussed degree distribution $P(k)$ peaked at $k = 2$ but with quite large values also for $k > 2$.

In synthesis, it is interesting to underline the good global topological connectivity properties of this network, which provides it with good robustness to random failures. With respect to the vulnerability assessment against targeted attacks, two different analyses have been performed: a topological and a reliability one.

In the topological analysis, only the connectivity pattern of the network has been considered and the vulnerability index has been evaluated with respect to the

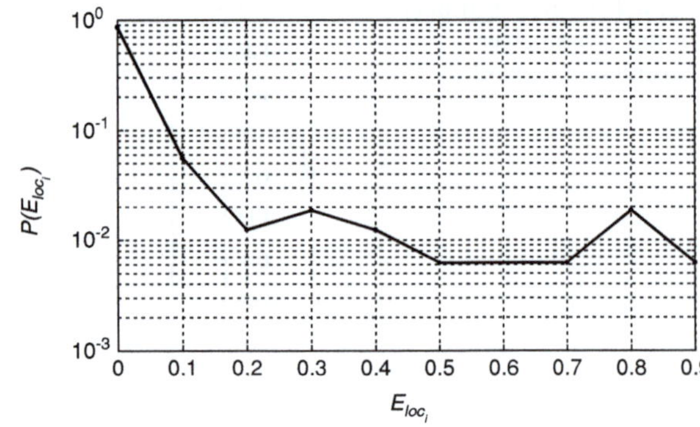

Fig. 6.5 Distribution of the local efficiencies for the subgraphs G_i

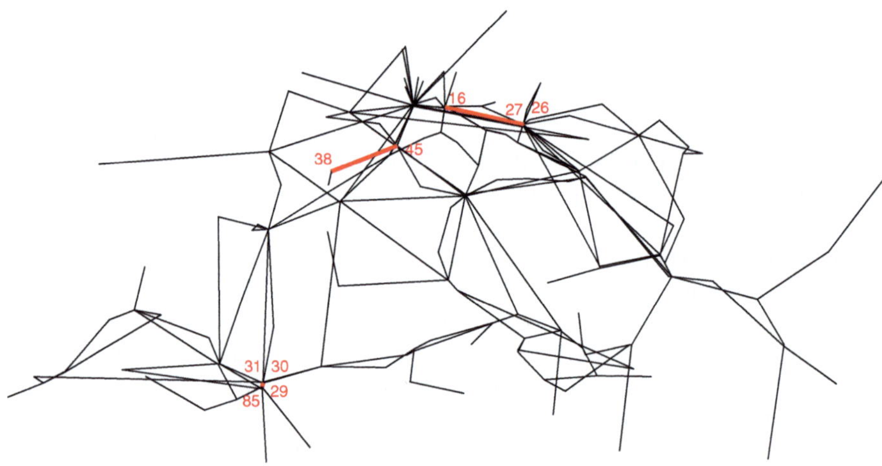

Fig. 6.6 The five most vulnerable lines according to the topological vulnerability evaluation (*numbers* mark starting and arriving node)

global topological efficiency. The five most vulnerable lines resulting from the topological vulnerability evaluation are displayed in Fig. 6.6.

In the reliability-weighted analysis for identifying vulnerabilities, both the connectivity pattern of the network and the reliability of each line have been considered. The same assumption on the line annual failure rates has been made. The vulnerability index has been evaluated with respect to the global reliability efficiency. The five most vulnerable lines according to the reliability vulnerability evaluation are displayed in Fig. 6.7.

Comparing the results in Figs. 6.6 and 6.7, one notices little difference between the vulnerable lines identified by the topological and reliability analyses. Indeed, the four lines ranked most vulnerable are actually the same in the two cases. This is possibly due to the crude assumption of equal failure rates which makes the reliability of a line only dependent on its length. This implies that the

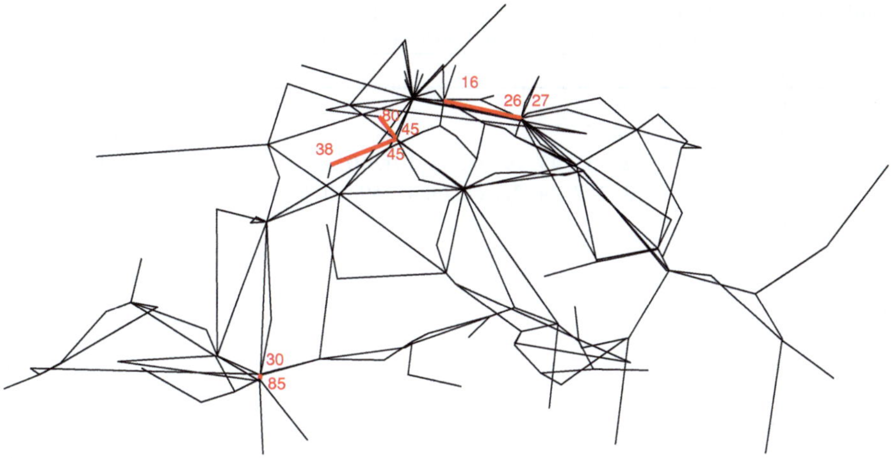

Fig. 6.7 The five most vulnerable lines according to the reliability vulnerability evaluation (*numbers* mark starting and arriving node)

Table 6.6 Summary of the criticality indicators rankings for the model of cascading failures (only the 24 most critical nodes are reported)

f_i	t_i	d_i	s_i
31	49	35	14
8	73	14	79
17	119	79	76
15	121	12	35
6	122	76	12
23	123	68	61
39	124	43	86
93	125	36	11
22	126	86	88
34	43	41	75
54	107	61	68
120	109	40	78
30	56	11	48
94	99	47	59
55	115	46	81
81	111	66	110
86	110	67	23
78	102	7	36
21	11	48	7
62	67	78	9
66	60	62	58
96	103	81	67
58	106	60	47
51	118	88	62

Table 6.7 Information, degree, closeness and betweenness centrality measure ranking for the network of Fig. 6.8 (only the 24 most central nodes are reported)

Rank	C^{I}	C^{D}	C^{C}	C^{B}
1	68	68	64	88
2	14	64	75	14
3	88	24, 35, 43, 79, 88, 92, 101, 103	79	75
4	119	2, 3, 7, 14, 21, 27, 28, 47, 59, 60, 67, 75, 81, 91	81	64
5	64		14	79
6	75		78	101
7	122		67	76
8	79		62	59
9	12		61	12
10	78, 110		63	110
11			76	61
12	101		88	102
13	59		41	98
14	123		71	68
15	76		60	71
16	47		65	83
17	43		59	84
18	24		68	40
19	35		73	67
20	61		82	35
21	121		86	78
22	71, 81, 98		83	60
23			80	81
24			40	107

reliability-weighted evaluation is very similar to a line length-weighted evaluation, with the weights being the lengths of the lines.

The topological, unweighted and weighted analyses are computationally very fast and easy to implement; in this sense, they are valuable tools for performing an initial screening of the vulnerabilities of the system so as to focus the directions of further detailed analysis by more sophisticated models.

The static character of the analysis cannot capture the dynamic behavior of the system, i.e. the dynamics of the loads and generators and their reconfigurations in the case of electrical transmission infrastructures. Moreover, the topological model which the system analysis relies upon does not take into account the load patterns in the system, so that the system vulnerabilities are identified based only on the connection patterns of the network. This is a limitation in the physical description of the system behavior since the load distribution on the overhead lines may not necessarily follow the topology of the system.

Finally, the Swiss grid is an open system with given energy flux boundary conditions with neighboring countries. These conditions should be taken into account as they may bring additional vulnerabilities to the network.

Fig. 6.8 The 380 kV Italian power transmission network (TERNA 2002; Rosato et al. 2007)

6.2.5.2 Example of Failure Cascade Modeling Application for a Single CI

Following the topological and the weighted analyses (Zio and Sansavini 2011a), the vulnerability analysis proceeds to the cascading failure assessment. As an example of application, the indicators of component criticality introduced in Sect. 6.4.3.1 have been computed for the topological network of the 380 kV Italian power transmission network (Fig. 6.8), considering cascades evolving according to the failure propagation model of Sect. 6.4.3. The 380 kV Italian power transmission network is a branch of a high voltage level transmission, which can be modeled as a network of $N = 127$ nodes connected by $K = 171$ links (TERNA 2002;

Rosato et al. 2007). Its topology is taken as reference but the failure propagation models applied to it have very little specifics to such a system; it is only used so to give concrete examples of the findings. In all simulations, the cascading failure evolution has been followed step by step, the relevant information collected and, eventually, the quantities Eqs. 6.21–6.24, have been computed.

Table 6.6 reports the results for the model relative to cascading failures presented in Sect. 6.2.3 due to targeted intentional attacks. The consequences of the disconnection of each individual node are estimated. This model describes a situation completely different from Sect. 6.4.4 since the load at a component is the total number of shortest paths passing through that component. Components 35, 14, 79, 12, and 76 turn out to be the most critical with respect to s_i and d_i whereas t_i gives an opposite ranking for the reason explained before.

The components ranked as the most critical according to f_i are now the ones with a small capacity, since initially they do not have many shortest paths passing through them but still are linked to highly connected components or lie along a direct path linking highly connected components (30 and 35 for critical component 31, LOM9; 7 and 2 for critical components 8, Casanova, and 6, Laboratorio Cesi; 24, 7, and 21 for critical components 17, LOM1, and 15, Musignano): while other components are failing and the highly connected components are still operating, the evolving shortest paths are directed through these critical components which subsequently fail due to their low capacity. Also, as expected, components of degree one do not participate to any cascade since they involve no shortest path transit.

The last two columns of Table 6.6 report a ranking with respect to the indicators d_i and s_i which is completely different compared to f_i. The most critical components form a path connecting the Northern area to the Tyrrhenian backbone, i.e. Vignole B.—La Spezia – Marginone—Poggio a Caiano; whenever this path is broken, i.e. either component 14 or 79 or 12 or 76 fails, the connectivity capability is shifted somewhere else in the network, i.e. in the Po River area, leading to an accruement of the cascade with further failures.

Overall, considering the different models, the ranking results of Table 6.6 are consistent with a physical analysis of the network system, indeed highlighting the components which most affect the failure spreading.

The results from the propagation of cascading failures from the network theory approach cannot be fully compared with those of the probabilistic modeling propagation logics since the former is a completely deterministic process, in which the components load is the number of shortest paths passing through it, while the latter involves a stochastic process of components loads distribution. In the case of network theory model the components ranked as the most critical according to f_i are those of small capacity, since initially they do not have many shortest paths passing through them but still are linked to highly connected components.

Table 6.7 reports the ranking of the individual network components according to the information (C^I), degree (C^D), closeness (C^C) and betweennes (C^B) centrality measures.

From the comparison between the results in Table 6.7 and those for the model relative to cascading failures (Table 6.6), it can be said that the betweenness and information centrality measures only partially account for the criticalities highlighted by the indicators d_i and s_i (node 14), since not only the centrality of a component is relevant but also the fact that it is a connecting central component (as do critical nodes 12 and 76 connected to central component 14 and critical node 79 connected to central component 75 in the network).

With respect to the indicators f_i, it can be said that the most critical components are those less connected, which lie along a direct path linking components with the highest degree centrality (35 and 28 for critical component 31; 3 and 7 for critical component 8; 24 and 7 for critical components 17 and 15; 2 and 7 for critical component 6).

The degree and betweenness centralities, which account for the number of connections pointing to a component and for the number of shortest paths passing through a component, respectively, appear to play a major role in identifying those network components which most contribute to the failure propagation process. The betweenness centrality measure only partially highlights those components which most contribute in determining large-sized failure cascades.

6.2.6 Conclusions

The methods of complex network theory can provide information useful for the vulnerability assessment of CIs, within a screening analysis that leads off to an adequate system understanding that cannot be superficial for the following detailed analysis. The analysis is supported by structural information provided by system owners, including the general understanding of main functionalities, interfaces, (inter-) dependencies, etc. The evaluation of the statistical indicators derived from the analysis provides indications of obvious vulnerabilities, e.g., structural or reliability bottlenecks, etc.

The topology-driven analysis of CI vulnerability is an essential part of the methodology, able to provide a reliable identification of the most critical connections, nodes, or areas in the network. The identification of the most vulnerable parts of a CI should always be complemented by a successive dynamical analysis of the cascading failure propagation process. Given the somewhat 'abstract' level of the modeling supporting the topological analysis, the results gained with respect to the vulnerable points (or lines) in the system ("first findings") may not be 'clear-cut' and major hidden vulnerabilities may still be expected. Then, to achieve a higher degree of accuracy, system understanding has to be further developed and more detailed information about the system and its operating environment may be needed. Special attention should be placed on interdependencies within or among systems. The re-assessment of simplifications made earlier may call for more sophisticated methods of their successive in-depth (detailed) analysis.

In conclusion, the specific goal of the application of complex network theory methods to the analysis of CIs is to fulfill two main objectives: helping (a) to identify preliminary vulnerabilities of critical infrastructures by topology-driven and dynamical analysis and (b) to guide and focus further detailed analyses of critical areas of the CIs.

6.3 Risk Analysis of Critical Infrastructures

Identifying and quantifying the vulnerability characteristics of critical infrastructures (CIs) is crucial for designing the adequate protections, mitigation, and emergency actions against failures. Methods of probabilistic risk assessment (PRA), also called quantitative risk assessment (QRA), for screening and prioritizing vulnerabilities can be helpful tools of CIs analysis.

The current PRA methodologies are successfully applied to the analysis of man–machine-environment systems with well-defined rigid boundaries, with single, well-specified targets of the hazard and for which historical data exist in support of uncertainty quantification models (Kröger 2005); examples are nuclear power facilities and spacecrafts. Currently, efforts are directed toward the study of the applicability of PRA methodologies to CIs (Grigg 2003; Garrick et al. 2004; Haimes and Horowitz 2004; Apostolakis and Lemon 2005).

6.3.1 Conceptual Outline

There are three elements that contribute to risk in a technological system: the sequences of failures events, their likelihood of occurrence, and the associated consequences they generate on specified targets (workers, public, environment, assets, etc.). The quantitative characterization of the triplet of elements composing risk, proposed by (Kaplan and Garrick 1981), provides an informative and operative definition for a systematic answer to the following three questions of risk analysis:

- Which sequences of undesirable events transform the hazard into an actual damage?
- What is the likelihood of occurrence?
- What are the consequences of each of these sequences?

Methods have been developed for answering these questions in complex technological systems, but they need some adaptation for dealing with the CIs technological and sociopolitical complexities.

PRA is a methodological framework that integrates deterministic and stochastic tools to carry out a systematic, exhaustive, and structured evaluation of the risk associated with every life cycle aspect of a complex engineered technological system, which may lead to undesired consequences triggered by an accident initiating event. Most current risk analysis methodologies begin with valuing targets followed by identifying potential failures, hazards and threats to which the system is exposed, and the related vulnerabilities; modeling, and computation are then used for determining risk and prioritizing the hazards, threats and vulnerabilities to devise appropriate countermeasures, usually on the basis of a costs/benefits analysis.

In a PRA, the risk connected to a particular accidental sequence is quantitatively characterized by two quantities: (I) the magnitude of the adverse consequences that may result from the operation of the system and (II) the likelihood of occurrence of the given adverse consequence(s). For example, the measure of consequence severity can be the number of members of the public that can be potentially injured or killed, in which case risk assessment becomes a powerful tool to quantify the public safety performance of the system; the likelihood of occurrence can be expressed as probability or frequency (i.e., the number of occurrences or the probability of occurrence per unit time).

If the severity of the consequences and their likelihood of occurrence are expressed qualitatively (e.g., through labels like high, medium, or low), the term called qualitative risk assessment is applied.

6.3.2 Modeling Techniques

6.3.2.1 Qualitative Assessment

Qualitative assessment allows the quick identification of potential hazards, threats and risks, and the related targets, assets, and resources which are vulnerable to these. In doing so, qualitative analysis verifies the effectiveness of the safety measures already implemented in the system and indicates those which could be useful to add. The goal is to guarantee an acceptable level of risk protection. Any use of calculations is kept fairly basic at this stage, and usually does not require the exact values of all the parameters in question.

The main outcome of the qualitative risk assessment is the prioritization of risks and corresponding areas for immediate action/protection and improvement. Risks are prioritized on high-medium-low categorical scales, based on the two criteria of severity of consequences and probability of their occurrence. The disadvantage of the qualitative analysis is embedded into its methodology which does not provide quantifiable evaluations neither of the magnitudes of the consequence, nor the probabilities of occurrence, which renders difficult a cost/benefit analysis of the safety measures already implemented or recommended.

Qualitative risk assessment bases the probability and consequences evaluation on expert judgments. Because qualitative analysis does not rely on actual numerical data, the analysis if far simpler and easier to implement and understand, but is affected by high degree of subjectivity.

Such assessment is recommended and often sufficient for simple systems, e.g. a single product, system, or process; the methodology is also useful for identifying and analyzing the threats and vulnerabilities of complex systems, and for a preliminary evaluation of the adequacy of the safety measures installed to mitigate the threats (Moore 2006).

An exemplary step-by-step procedure for general qualitative risk assessment is presented in the following (Moore 2006). The illustration is customized to manage the wide scope of parameters which characterize CIs.

Step 1—system characterization: The system characterization involves (I) the analysis of the information on the technical details of the system facilities, (II) the identification of the potential critical assets, (III) the identification of the hazards, threats, and the consequences of concern, and (IV) the identification of existing safety measures and protection layers.

Step 2—hazard and threat assessment: The consideration of possible hazards and threats should include accidents, attacks, and failures (Chap. 4). The assessment should be based on local, regional, or national information, where available. For malevolent targeted attacks, the determination of the attractiveness of the system as a target from the adversary's perspective is to be considered.

Step 3—vulnerability analysis: The vulnerability analysis involves pairing up target assets with hazards and threats to identify potential vulnerabilities. This entails the identification of the existing countermeasures and their level of effectiveness in reducing the vulnerabilities. The degree of vulnerability is evaluated either (I) by the formulation of risk scenarios or (II) on the basis of asset protections. Higher consequence and attractiveness ranking typically call for the application of a scenario-based approach of vulnerability analysis. This involves the assignment of risk rankings to the scenarios developed. On the contrary, if the asset-based approach is used, the determination of the consequences on the asset and its attractiveness may be enough to perform a target ranking value and to identify a standard protection set for that target level.

Step 4—risk assessment: The risk assessment determines the risk to the system combining the expected effects on each critical asset. For malevolent attacks, likelihood is determined by a team of experts after considering the attractiveness of the targeted assets assessed in step 2, the degree of threats assessed in step 2, and the degree of vulnerability identified in step 3.

Step 5—safety countermeasures analysis: Based on the vulnerabilities identified and the risk that the layers of protection are breached, appropriate enhancements to the safety countermeasures are recommended to reduce vulnerability of the system, following the typical doctrines of deter, detect, delay, respond, mitigate, and possibly prevent. Some of the guiding factors to be considered in these recommendations are:

- The decrease of the probability of successful failure/attack
- The degree of risk reduction
- The reliability and maintainability of the options
- The capabilities and effectiveness of mitigation options
- The costs and feasibility of mitigation options

Based on these factors, the countermeasure options should be ranked to evaluate their effectiveness, and prioritized to assist decision making for implementing safety enhancements. In case of terrorist threats, the consequences of a security event at a facility are generally expressed in terms of the degree of acute health effects (e.g., fatality, injury), property damage, environmental effects, etc. This definition of consequences is the same as that used for accidental releases in industrial plants, and proves appropriate also for security-related events; the key difference is that security-related events often involve effects that are more severe than those related to accidental risk.

6.3.2.2 Quantitative Assessment

6.3.2.2.1 Guiding Principles

Although the literature suggests many different risk assessment methodologies, in fact the differences are primarily in scope, application, boundary conditions, degree of quantification and quality. Like many other scientifically based methodologies, quantitative risk assessment is founded on relatively few basic principles.

Guiding principles for scenario-based risk assessment within the framework of vulnerability analysis of CIs (Chap. 4) are listed below (Garrick et al. 2004):

- The quantitative expression of risk should be in the form of a structured set of scenarios, each having a corresponding likelihood and consequence.
- The set of scenarios must be complete in the sense that all of the important contributors to risk are included.
- The scenarios must be quantified in terms of clearly defined risk measures, must be realistic, and must account for the involved uncertainties.
- Each scenario should be characterized by a sequence of events, starting with the initiating event that upsets an otherwise successfully operating system and proceeding through a series of subsequent events to the end-state. The identification of the initiating events must be based on a comprehensive hazard and threat assessment.
- Each scenario must accommodate combined events, including primary and diversionary events.
- The end-states i.e. the undesired consequences. must reflect immediate, cascading, and collateral consequences (or levels of damage).
- Uncertainties relative to individual and aggregated events must be quantified on the basis of the available evidence, with the appropriate mathematical techniques.
- The results must be ranked as to their contribution to risk in order of importance and must be presented in a way that supports decision making.

6.3.2.2.2 Implementation of the Principles

Adherence to these principles may be achieved through the following six-step process (Garrick et al. 2004):

(1) Defining the system being analyzed in terms of what constitutes normal operation and points of vulnerability, to serve as a reference point.
(2) Identifying and characterizing the 'sources of danger', that is, the hazards (e.g. stored energy, toxic substances, hazardous materials, acts of nature, sabotage, terrorism, equipment failure, combinations of each, etc.).
(3) Developing "what-can-go-wrong" scenarios to establish levels of damage and consequences while identifying points of vulnerability.
(4) Adopting risk metrics that reflect the likelihoods of different scenarios and quantify the consequences of the scenarios based on the totality of relevant evidence.
(5) Assembling the scenarios according to damage levels, and casting the results into the appropriate risk curves and risk priorities.
(6) Interpreting the results to guide the risk-management process.

Step 1—defining the system: The purpose of the first step is to understand how the system works to allow the identification of departures from normal, successful operation. Once the system is understood, vulnerabilities that require special analysis can be assessed.

The following aspects, specific to CIs, must be taken into account:

- In an increasingly interconnected world, technology-based systems and networks are becoming more and more interdependent. Therefore, the geographical, logical and functional situation should be carefully investigated: boundaries of the system, environmental interactions, external interactions and associated constraints, e.g. interdependencies with other CIs.
- The composition of the system: identification of all subsystems, identification of internal interactions with associated constraints.
- The potential of far-reaching, cascading effects of failure of one system on other systems and on society as a whole: this entails that the dynamics of the system behavior should be analyzed as well.

The relevant infrastructures must be modeled in such a way that the scenarios can be readily identified. In some cases, such as electrical transmission systems, railway transportation infrastructures, water-supply networks, the CIs can be modeled using networks to take advantage of existing mathematical tools of network analysis.

Step 2—characterizing hazards and threats: The identification of the initiating events (IEs), i.e. hazards and threats to the system, represents a major part of the risk analysis since many of the subsequent steps depend on the type of the considered hazards and threat, e.g. accident, attack or failure (Chap. 4), and on its intensity.

In the case of IEs such as accident events and other phenomena with inherently random nature, in order to focus on the most important initiators while screening out the unimportant ones, several methods have been proposed in the literature, e.g. checklists, What-If Analysis, Failure Modes and Effects Analysis (FMEA) and Master Logic Diagrams (MLDs). While random IEs can be more easily identified, difficulties appear when terrorism or malevolent acts are considered, due to problems in determining the likelihood of a successful attack. The assessment of the likelihood that a terrorist attack will occur requires information on the intent, capability, and resources to carry out the attack (Koonce et al. 2008). Given that a group possesses these traits, determining the point, or points, of attack requires knowledge of the goals, beliefs, and desires of the group. The probability of the attack being successful depends upon the quality of countermeasures in place to deter or combat the attack (Paté-Cornell and Guikema 2002). For these reasons, threats of appropriate levels can be assumed for the analysis and the evaluation of the likelihood of attack is left to intelligence agencies (Apostolakis and Lemon 2005).

Step 3—constructing scenarios: Scenario development follows a structured format that answers two of the risk triplet questions: what can go wrong? and what the consequences would be? A variety of logic and analytical tools are used to develop scenarios, such as FMEA, state space diagrams, event trees. The state space diagram is a logic model depicting the various states of a system and the paths along which the system can transfer from one state to another. The diagram can be represented by a set of simultaneous differential equations describing how the probability of the states changes with time. State space analysis is a useful tool for displaying important elements of the evolving scenario. Event trees are common methods used for scenario development and involve going forward from an initial disturbance of the system. These methods, in combination with fault trees, can be used to construct an encompassing set of risk scenarios. Obviously, a comprehensive examination of system vulnerabilities might identify a great number of possible threats for a particular set of consequences or damage levels.

A method that can be used to determine the physical consequences resulting from the failure of the components in an accident scenario is the actual simulation of the behavior of the systems involved in the scenario. This can be a difficult task in the majority of the cases that imply CIs; attempts have been made for analyzing an electrical transmission network by a load flow simulation (Koonce et al. 2008).

A system is exposed to different hazards and threats according to its functionalities, its global environment and of course geopolitics context. Thus, the building of scenarios should take into account all these factors. Piwowar et al. (2009) suggest using different methods according to the criticality of the system. By methods such as Bayesian analysis (Hudson et al. 2002) or other structured, quantitative approaches (event or causal trees, bow-tie diagrams, etc.), a large part of accidents and malevolence act scenarios can be identified and protection must be efficiently defined in order to anticipate and protect the system. On the other hand, well-prepared and organized terrorist organizations might carry out a vulnerability assessment themselves and arrive at the same conclusions. Game-theoretical paradigms can be applied in these cases (Paté-Cornell and Guikema 2002; Kardes 2005).

Non-probabilistic methodologies can also be employed, such as possibility theory based on belief/plausibility measures (Darby 2006), fuzzy logic (Ross 2004), meta-heuristics (Hansen et al. 2007; Piwowar et al. 2009), and robust decision making (Lempert 2004).

Step 4—risk assessment: Risk assessment encompasses a number of techniques, e.g. fault trees, event trees, Monte Carlo simulations, influence diagrams, multiple attribute theory, common-cause failures models, sensitivity analyses, value trees, etc., but the underlying basic principles are few. The principles focus on the development of scenarios describing how the system under study is supposed to work and scenarios indicating how the system can be made to fail. The step implies a systematic evaluation of causes, frequencies and consequences of each undesired condition. The likelihood of events in the scenario must be linked to the supporting evidence. The indicator selected for measuring risk is based on the success rate of different levels of damage and therefore on the method chosen for constructing the scenarios.

The determination of the conditional probabilities for the quantification of the event trees representative of the accident or attack scenarios are based on calculations and analyses of varying complexity, by deterministic and probabilistic codes, supported by factual data and engineering judgment, and system analysis codes to assess the availability of protective systems. The fault tree methodology is an example of methodology often used in practice. It is an effect-cause representation of logic that works backward from the undesirable end state.

An exemplary methodology for the evaluation of the consequences of an undesired condition is the value tree (Koonce et al. 2008), based on multi-attribute utility theory (MAUT), which provides a hierarchical view of the impact each failure scenario may have on the stakeholders involved on the consequences of the accident. The value tree consists of three levels in which the top level is the overall impact, or value, of a failure scenario; the second level breaks this overall impact into broad categories called impact categories; the impact categories are further reduced in the third level to specific aspects, called performance measures, which specifically describe the various ways consequences result in impacts to the stakeholders. Each performance measure is divided into various levels of impact called the constructed scales. The levels of the constructed scales represent the amount of impact the physical consequences have on the stakeholders through each performance measure. The levels for each constructed scale range from no impact to complete impact to the performance measure.

Step 5—assembly: Once the individual scenarios have been assessed, they can be assembled into risk measures, such as the frequency–consequence curve. This is a matter of combining all scenarios that terminate in a specific damage category. If the risk measure is a numerical variable, such as fatalities, injuries, or dollars, then the process also involves arranging the scenarios in order of increasing damage and cumulating the probabilities from bottom to top. Showing different levels of damage, such as the risk of varying injuries or fatalities, requires a different type of presentation. The most common form is the classical risk curve, also known as the frequency-of-exceedance curve, or the complementary-cumulative distribution function (CCDF). This curve is constructed by ordering the scenarios by increasing levels of damage and cumulating the probabilities from the bottom up in the ordered set against the different damage levels (Fig. 6.9).

Sequence	Probability	Consequence
s_1	p_1	x_1
s_2	p_2	x_2
...	...	...
s_n	p_n	x_n

(a)

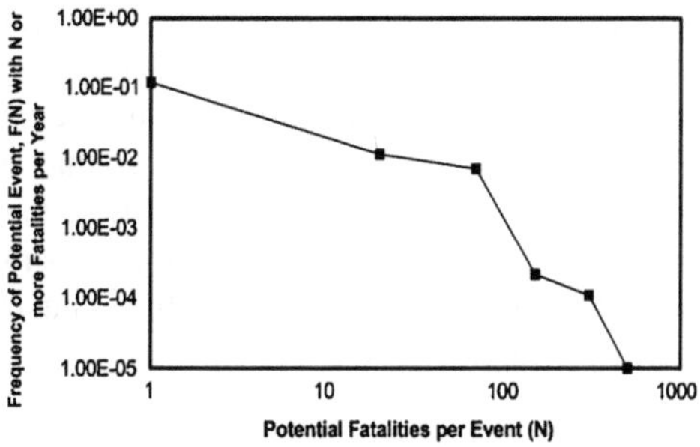

(b)

Fig. 6.9 **a** Risk as a list of triplets and **b** example of frequency–consequence curve (Zio 2007b)

Depending on the method chosen in steps 3 and 4, different risk measures can be used for the identification of vulnerabilities and ranking of the elements. Apostolakis and Lemon (2005) propose a performance index (PI) similar to expected utility, calculated for each minimal cut set (MCS) identified through the methodology. The assessment determines the susceptibility of each MCS to this level of threat by looking at its accessibility and security measures. The susceptibility of each MCS is then combined with its value for ranking. The result is a ranking which prioritizes MCS having both high susceptibilities and high values.

Patterson and Apostolakis (2007) generalize the performance measures to include the stakeholder values. Then, each infrastructure is analyzed independently to assign to each location a value of importance (geographic valued worth, GVW); the GVWs from each infrastructure for a given location are summed to determine the location's overall GVW; these GVWs are used to rank the various locations.

Step 6—interpret the results: An important output of a risk assessment is the identification of the dominant contributors to the risk, for effective risk management. The vulnerabilities and their ranking according to potential impact are a useful input to risk-informed decision making.

Most risk assessment software packages contain algorithms for ranking the importance of contributors to a risk metric when the results are obtained under the form of risk curves.

Also, the ranking can be achieved by using MAUT. The use of MAUT allows to treat risk as a multiattribute concept, and to have a basis for ranking the vulnerabilities that include the values of the decision maker and the relevant stakeholders.

6.3.2.2.3 Cascading Failures

The cascading failure propagation is accounted for in steps 3 and 4 for the construction of scenarios. In the extended methodology of Koonce et al. (2008), this is done during the power flow simulation. The load flow simulation uses a quasi-steady state step-in-time to determine the effects an initial, single-component failure has on the entire infrastructure. This step-in-time simulation has the ability to identify components in the grid that experience conditions outside of their limits, e.g., transmission lines that experience over-current conditions. These components are then tripped off-line causing a cascading effect, which can result in the initial, single-component failure causing additional load shed than a normal stability analysis would conclude.

6.3.2.2.4 (Inter)dependencies Identification and Modeling

As stated in Chap. 2, dependencies and interdependencies identification and modeling represents a difficult task when analyzing CIs and can be treated in step 1, i.e. system definition.

The consequences of dependence between failure-concurrent events may be severe in case they affect redundant components or systems whose operations must take place simultaneously or within a short time interval. These types of dependent failures are referred to as common-cause failures and should be explicitly represented in the logic models identifying accident scenarios, e.g. in the event trees and fault trees. The quantification of common-cause failures is challenging, mainly because observations and available data are normally scarce.

The modeling of interdependent CIs is treated in Apostolakis and Lemon (2005), Michaud and Apostolakis (2006), Patterson and Apostolakis (2007), and Piwowar et al. (2009). Piwowar et al. (2009) gives a representation of the system in critical layers to take into account that some under-systems could be linked by feedback loops or could belong to causal modes. In Apostolakis and Lemon (2005), interdependent infrastructures are accounted for in the natural network modeling through the introduction of an additional vertex which models the interconnection of the two networks. In Michaud and Apostolakis (2006), the durations of system failures are included to capture the time dependence of the consequences resulting from failures in interdependent infrastructures.

6.3.2.2.5 Interdiction/Protection Optimization

According to the results of the quantitative assessment, the decision makers can substantially contribute to the improvement of safety and security protective systems. Several ways to strengthen these systems can be suggested, that would permit stakeholders to have a wide choice on the possible improvement strategies according to a cost/benefit analysis. Stakeholders can decide which strategy is the most pertinent for the infrastructure according to the most undesirable accident scenarios. Based on these risk-based insights, systematic examination of successive improvements would continue until an acceptable level of overall risk is achieved.

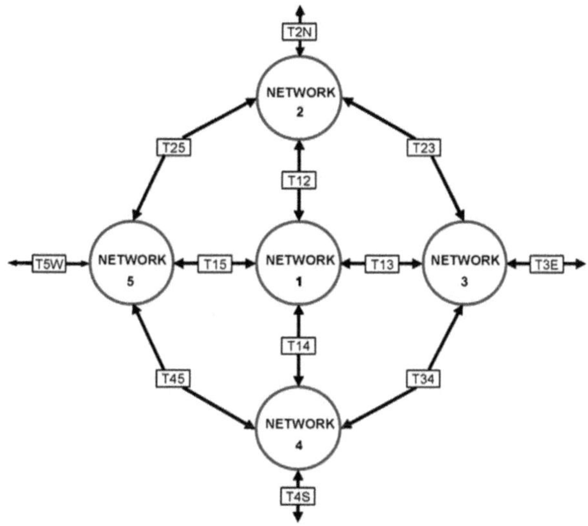

Fig. 6.10 Sample regional grid (Garrick et al. 2004)

6.3.3 Exemplary Application: Assessing Vulnerabilities of a Terrorist Attack on a Power Grid

The example (Garrick et al. 2004) involves a risk assessment of a physical attack on a hypothetical electric power grid, following the step-by-step procedure previously presented in Sect. 6.3.2.2.

The example shows how CIs can be analyzed to highlight their vulnerabilities and provide a basis for taking corrective actions to avert or mitigate the consequences of a terrorist attack. The risk assessment of the sample electrical grid leads to specific safety recommendations that could not have been easily deduced without an analytical approach.

Figure 6.10 represents a major region in the national electric power grid; each network corresponds to a large metropolitan area. Networks are interconnected to form a regional grid. In Fig. 6.10, network 1 is interconnected with four neighboring networks through connections T12–T15. These interconnections form the transmission network and are extra-high voltage (EHV) transmission lines that provide the major pathways for power flow throughout the region.

Figure 6.11 shows an expanded view of network 1, which is the electrical power distribution network for a particular city. The distribution system forms a network of generators, substations, and major overhead lines. The network 1 has five generating stations (G1–G5) and four major substations (S1–S4) that distribute power to consumers. Generation and distribution power flows in each network are typically coordinated through centralized operations from a control facility.

Once the system is defined, the associated hazards are identified. In this example, the source of danger is defined as a potential terrorist action. The considered scenario is the attack to the electrical grid and six potential end states are

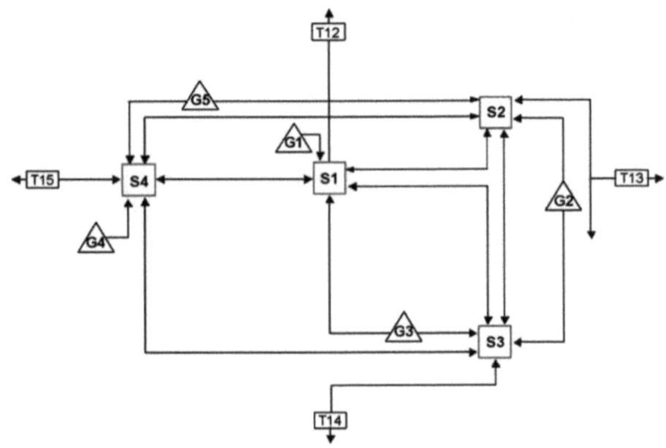

Fig. 6.11 Generators and distribution substations in network 1 (Garrick et al. 2004)

defined and linked to corresponding initiating events through the scenario development process:

Damage level 0 (no damage):	No significant network or regional power outages
Damage level 1:	Transient outage to network 1
Damage level 2:	Transient outage to the region (and network 1)
Damage level 3:	Long-term outage to network 1
Damage level 4:	Long-term outage to network 1 and transient outage to the region
Damage level 5:	Long-term outage to the region (and network 1)

As detailed in step 3, a variety of logical and analytical tools are used to develop scenarios. For the scenario involving the attack on network 1, numerous physical methods could be used to damage equipment at each substation with varying degrees of damage to the network and the region. For example, carbon fibers, Mylar strips, or other contaminants could be sprayed over buses and transformers to cause severe short circuits. Explosives could be used to destroy key transformers, circuit breakers, and bus sections. Attackers could also damage circuit breaker controls at substation operating panels.

To model this scenario, it will be assumed that a threat assessment uncovered a high likelihood that detailed information about the electrical grid has been made available to terrorists. To illustrate the method, substation S1 is analyzed first because it controls the whole output from generating station G1, part of the output from generating station G3, and, most important, the termination of the key regional interconnection T12.

Figure 6.12 shows the systematic process used to develop the attack scenarios and to assign their consequences to the damage levels. Branches may be added to account for other protective barriers in each system. The purpose is to comprehensively identify vulnerabilities, for guiding decision making. A full-scale risk assessment would detail the effects from attacks on each substation, as well as multiple substations at once.

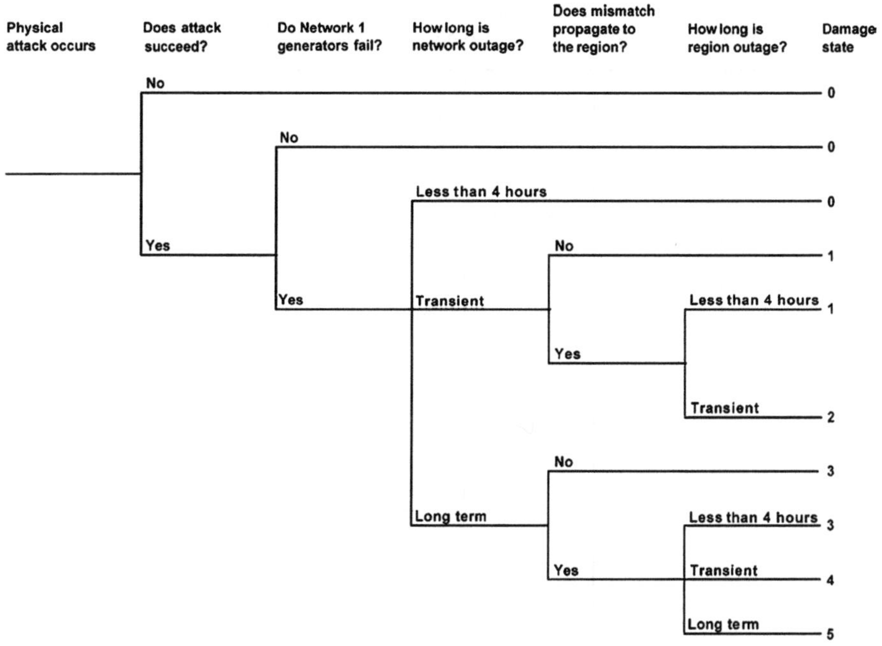

Fig. 6.12 Attack scenario (Garrick et al. 2004)

The event tree in Fig. 6.13 presents a more detailed analysis of the network 1 top event. The expanded logic includes more details about the attacks to the three critical substations in network 1 and the corresponding likelihoods of a long-term network power outage.

The scope and definition of each top event listed in Fig. 6.13 are summarized below.

The SUB S1 top event represents the success of destroying sufficient equipment in substation S1 to disable its power generation and transmission interconnections. The horizontal path from the SUB S1 top event occurs if the attackers do not achieve their goal and the substation may be partially damaged, or power is only temporarily disrupted. In this scenario, the damage is not sufficient to incapacitate the major interconnections for more than 24 h. The failure path from the SUB S1 top event (the vertical path in the event tree) occurs if the attackers cause enough damage to disable substation S1 for an extended time.

The SUB S2 top event is similar to the SUB S1 top event. It represents the success rate for the attackers destroying enough equipment in substation S2 to disable its power generation and transmission interconnections.

The SUB S3 top event is similar to the SUB S1 top event. It represents the success of destroying enough equipment in substation S3 to disable its power generation and transmission interconnections.

The NET top event represents the conditional success of each level of substation damage causing an extended power outage throughout network 1.

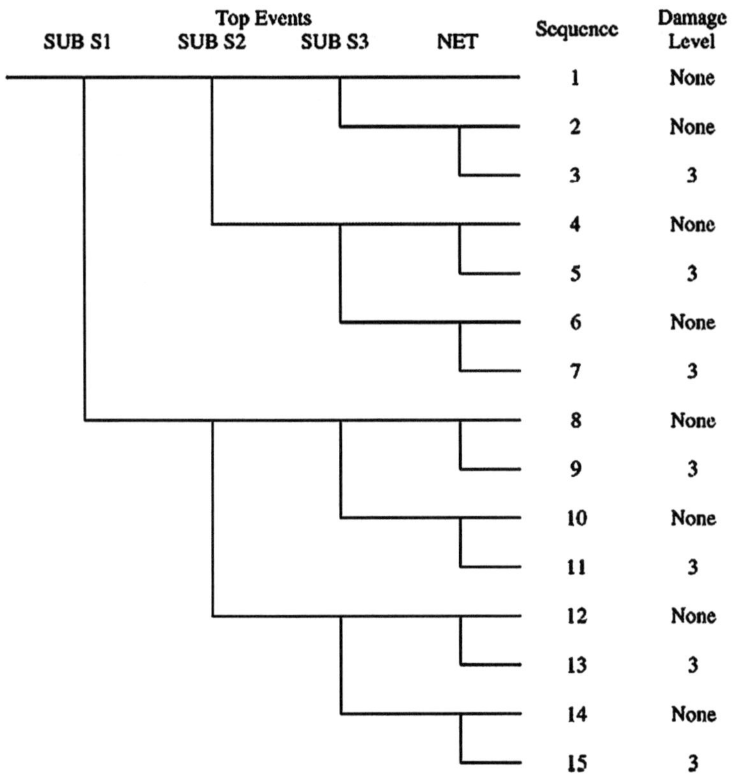

Fig. 6.13 Event tree for network 1 (Garrick et al. 2004)

This success depends on the specific combination of damaged substations, their generation and transmission interconnections, and the network loading conditions at the time of the attack. The success rate for a consequential failure of the NET top event is different for each combination of damage conditions to substations.

Sequence 1 in the event tree occurs if the attackers do not cause enough damage to incapacitate any of the three critical substations. Short-term, localized power disruptions may occur in some areas, but the outages are not of sufficient severity or duration to satisfy the critical damage criteria. The success path from the NET top event also occurs if the attacks do not inflict enough damage to cause widespread extended outages throughout the network. Thus, sequences 1, 2, 4, 6, 8, 10, 12, and 14 end in successful power supply.

Sequence 3 in the event tree occurs, if the damage to substation S3 is severe enough to cause prolonged power outages throughout a large portion of network 1. The failure path from the NET top event occurs whenever the achieved level of substation damage is severe enough to cause widespread extended outages throughout the network. This condition occurs in sequences 3, 5, 7, 9, 11, 13 and 15 and it is equivalent to damage level 3.

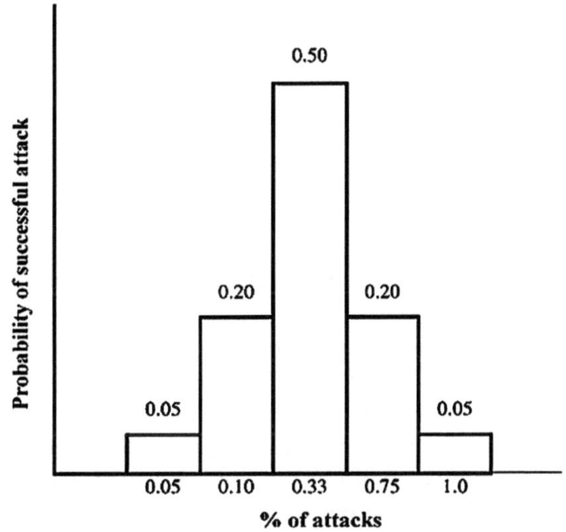

Fig. 6.14 Success rate of an attack on substation S1 (Garrick et al. 2004)

Top events SUB S1, SUB S2, and SUB S3 in Fig. 6.13 represent the likelihood that attackers would destroy enough equipment in each substation to disable its generating supplies and transmission interconnections. In this model, each substation is assigned a different vulnerability to attack:

Substation S1: It is assumed that substation S1 is located in an urban environment and is the most heavily protected of the three substations. It may be surrounded by protective walls, may be continually manned by utility personnel, and may be checked by local police during their normal neighborhood surveillance patrols.

Substation S2: It is assumed that substation S2 is located in a suburban or partially rural environment and is the least protected of the three substations. It may be surrounded by a chain link fence, may not be manned, and may not be subject to routine surveillance by local police.

Substation S3: It is assumed that substation S3 is located in an urban environment but is only partially protected. For example, it may be surrounded by protective walls and checked by local police during their normal neighborhood surveillance patrols, but it may not be continually manned.

A probability distribution gives the likelihoods that attackers could successfully enter each substation and cause extensive damage to critical transformers, circuit breakers, buses, and controls; the histogram in Fig. 6.14 is an example which could apply to substation S1; for example, it shows that there is a 5% probability that the attackers would succeed in 5% of their attacks on substation S1 (i.e. that 1 of 20 attacks would be successful). The mean likelihood of a successful attack on

Table 6.8 Estimated success rate of an attack (Garrick et al. 2004)

Substation	Probability					Mean
	0.05	0.20	0.50	0.20	0.05	
S1	0.05	0.10	0.33	0.75	1.0	0.39
S2	0.75	0.85	0.90	0.95	1.0	0.90
S3	0.10	0.30	0.50	0.80	1.0	0.53

substation S1 is approximately 0.39 (i.e. approximately 10 of 26 attacks would be successful).

The estimates underlying such histogram are obviously not derived from detailed models of specific attack scenarios or from a detailed evaluation of the specific substation vulnerability to each attack; rather, they are developed based on information from experts familiar with potential attack strategies, resources, and specific vulnerabilities of the target. If these preliminary results showed that attacks on substation S1 were potentially important to one of the undesired damage levels, more extensive and quantitative analyses would be justified.

Table 6.8 summarizes the estimates of a successful attack on each substation, considering its specific vulnerabilities. For example, concerning the first column there is a probability of 5% that attackers would succeed (the ordinate of the histogram of Fig. 6.14):

In 5% of their attacks on substation S1
In 75% of their attacks on substation S2
In 10% of their attacks on substation S3

An evaluation of the NET top event in Fig. 6.13 accounts for the conditional likelihood that each level of substation damage would cause an extended power outage throughout network 1. The description of the network indicates that substation S1 is the most important one because it controls the whole output from generating station G1, part of the output from generating station G3, and the termination of key regional interconnection T12; substation S2 is next in importance because it contains the connections from generating stations G2 and G5, which are not directly connected to substation S1; substation S3 is the least important of the three critical substations.

The network is designed to withstand the complete loss of any one substation under normal loading conditions. However, under severe loading conditions, attack-initiated faults might cascade to other substations and generating units. Therefore, the models for the NET top event must assign a likelihood of network failure after any combination of substations is damaged. In this simplified example, these conditional likelihoods are expressed as in Table 6.9. In a more detailed analysis, additional supporting information for these estimates could be derived from dynamic load-flow simulations, models of system response, interviews with network operations personnel, etc.

In this example, the focus was on network 1, which was determined to be the most vulnerable to long-term outages. The overall vulnerability of network 1 was

Table 6.9 Conditional success rate for network 1 failure (Garrick et al. 2004)

Damaged substations	Probability					Mean
	0.05	0.2	0.5	0.2	0.05	
S1	0.05	0.1	0.25	0.5	0.75	0.29
S2	0.01	0.1	0.15	0.25	0.5	0.17
S3	0.01	0.05	0.1	0.2	0.25	0.11
S1 and S2	0.25	0.5	0.75	0.9	1.0	0.72
S1 and S3	0.1	0.25	0.5	0.75	1.0	0.51
S2 and S3	0.05	0.1	0.25	0.5	1.0	0.3
S1, S2, and S3	0.9	0.92	0.95	0.98	1.0	0.95

Table 6.10 Risk of coordinated attack (Garrick et al. 2004)

Damage to substations	Likelihood of success	Fraction of total damage level 3 (%)
S1 and S2 and S3	1.74×10^{-1}	39.4
S1 and S2	1.17×10^{-1}	26.5
S2 and S3	8.72×10^{-2}	19.8
S2	4.42×10^{-2}	10.0
S1 and S3	1.02×10^{-2}	2.3
S1	5.20×10^{-3}	1.2
S3	3.61×10^{-3}	0.8

most strongly determined by the relatively high vulnerability of substation S2, in spite of the fact that this substation was not as important to the network electrical stability as the other more secure substations.

Table 6.10 shows that successful attacks on substation S1 would contribute to approximately 69% of damage level 3; successful attacks on substation S2 would contribute to approximately 96% of damage level 3; successful attacks on substation S3 would contribute to approximately 62% of damage level 3. These so-called fractional-importance measures are simply the sum of scenarios that include damage to each substation, divided by the total number of scenarios. Therefore, the overall vulnerability of network 1 is mainly determined by the relatively high vulnerability of substation S2, even though this substation was not individually as important to the network power generation and to the distribution network as the more secure substation S1.

For coordinated physical attacks, one very clear action to consider would be to improve the security of substation S2, which was identified as the principal contributor to long-term outages for network 1. This priority might not have been evident without performing the integrated assessment of the vulnerabilities and the potential consequences of the failure of each substation.

It is also very clear from Table 6.10 that attacks on multiple substations would greatly increase the likelihood of network 1 failure. Thus, substation security in general would be an important consideration in improving the security of the regional grid. The relative importance of the subsystems to the overall vulnerability would not be apparent without an integrated model that systematically evaluates each contribution to the damage.

6.3.4 Conclusions

PRA is a systematic and comprehensive methodology to evaluate risks associated with a complex engineered technological entity and therefore a potentially useful tool to assess vulnerabilities in CIs. PRA methods for vulnerability assessment of CIs can be classified in two categories, namely, qualitative and quantitative.

Several screening methodologies that identify and prioritize vulnerabilities and critical locations in infrastructures have been developed in the past years and their general approaches cover a wide area of applications.

The qualitative assessment of CI vulnerabilities by expert judgment and tabular methods can be useful to identify points of action to prevent or mitigate the effects of a potential accident or attack. To be effective, it must take into account several parameters to be as exhaustive and efficient as possible. As a consequence, the process is quite long and complex but it appears of primary importance in dealing with hazards and threats coming also from the outside of the CI.

The quantitative approach complements and extends the qualitative methodology by introducing the comparison of the net effects of multiple hazards and threats, both in terms of probabilities and consequences, and the combination of dependent factors. Therefore, it may help in reducing investments aiming at the prevention of events which have already happened without eliminating the possibility of improving defense against known types of accidents and attacks.

The quantitative developed models become key elements for a systematic risk management to evaluate the effectiveness of proposed improvements against CI vulnerabilities. The updated analysis results display the corresponding changes in the CI risk. Based on the risk-based insights obtained from PRA, systematic examination of successive improvements would continue until an acceptable level of overall risk is achieved.

In both qualitative and quantitative methodologies, expert judgment is required. 'Subject matter experts' are employed to quantify the effects of the human component on the vulnerability of CI. The results of the analysis are also stakeholder dependent. Different stakeholders, such as a federal agency, could include additional performance measures and discard some of those included in the analysis. However, the methodology would be unchanged and result in a component ranking appropriate for the new stakeholder views. The consequences associated with these failures are mainly determined by looking at the type and number of customers affected. Stakeholder input is used to evaluate the relative importance of the consequences.

Nevertheless, investigating risks and vulnerabilities for CIs has to go beyond the usual cause–consequence analysis when dealing with strong interdependent systems (IRGC 2006). Indeed, the behavior of a complex system cannot be described as the sum of the behavior of its individual elements. This renders questionable the suitability of classical risk analysis methods, e.g. fault tree analysis, which are typically founded on a decomposition of the system into subsystems and basic elements and their subsequent recomposition for quantification. Furthermore, predefined causal chains, e.g. identified by event tree analysis, seem inappropriate to identify the hidden risks and vulnerabilities emerging in a CI. On the other hand, simulation techniques may be recommended as 'scenario generators', but their computational cost may be excessive on real-size systems.

6.4 Probabilistic Modeling of Cascading Failures Dynamics

6.4.1 Introduction

Probabilistic dynamics modeling offers a valuable framework for quantitatively describing the process of cascading failures in CIs. The relevance of such modeling framework is due to the fact that failure cascading is the typical mechanism for significantly damaging accidents in CIs. For example in the Internet, failures requiring a rerouting of the information traffic eventually lead to an avalanche of overloads on routers that are not equipped for significant extra traffic. Also cascades in social and economic systems are similar to cascading failures in physical CIs in that initial failures create operating and environmental conditions that increase the likelihood of subsequent failures, leading to the emergence of a pattern of accident evolution that is extremely difficult to predict, even when the properties of the individual components are well understood.

6.4.2 Conceptual Outline

To understand the dynamics of failures spreading in complex networks, two main aspects must be considered. On the one hand, the topology of the underlying network plays a role; on the other hand, the characteristics of the network structure must be linked to the dynamic mechanisms of interaction and the possible spreading of failures that these can generate. In random sparse networks, this can be achieved by considering interacting agents whose decisions are determined by the actions of their neighbors (Watts 2002). For example, this is what economists call *information cascades* during which individuals in a population exhibit herd-like behavior because they are making decisions based on the actions of other individuals rather than relying on their own information about the problem. Within the formalism of network theory (Sect. 6.2), the *information* network structure is represented by *vertices* (or *nodes*) joined in a graph by *edges* (or *arcs*); p_k, the probability with which each node is connected to k neighbors, is the *degree distribution* of the graph; and $\langle k \rangle = z$ is the *average degree* (*coordination number*). Because many decisions are inherently costly, requiring commitment of time or resources, the relevant decision function frequently exhibits a strong *threshold* nature: the nodes of the network display inertia in switching states, but once their individual threshold has been reached, the action of even a single neighbor set of incoming signals can tip them from one state to another.

Two quantities are of interest in the analysis of the failure cascades dynamics: the probability that a *global cascade* will be triggered by a single node (or small set of nodes), where a global cascade is formally defined as cascade that occupies a finite fraction of an infinite network, and the expected size of a global cascade once it is triggered.

When regarded more generally as a change of state—not just a decision—the model belongs to a larger class of spreading problems that includes models of failures in engineered systems: in the following, power transmission networks are taken as reference dynamics as failure mechanisms of large electric blackouts provide a natural context for the cascading failure modeling needs, challenges and characteristics.

6.4.3 Cascade Mechanisms in Power Systems

Bulk electrical power transmission systems are complex networks made of large numbers of components that interact in various ways. When component operating limits are exceeded, protections act and the component "fails" in the sense of being no longer available to transmit power. Components can also fail in the sense of misoperation or damage due, for example to aging, fire, weather, poor maintenance or incorrect design or operating settings. A relay may have an undetected defect that remains dormant until abnormal operating conditions are reached: this is often referred to as a hidden failure (undetectable during normal operation but will be exposed as a direct consequence of other system disturbances, which might cause a relay system to incorrectly and inappropriately disconnect circuit elements) (Chen et al. 2005). These cascading misoperations are what leads to major system disturbances.

In any case, the failure causes a transient in which the power flow of the component is redistributed to other components according to circuit laws, and subsequently redistributed again according to automatic and manual control actions. The transients and readjustments of the system can be local in effect or can involve components far away, so that a component disconnection or failure can effectively increase the loading of many other components throughout the network. The interactions involved are diverse and include deviations in power flows, frequency and voltage as well as operation or misoperation of protection devices, controls, operator procedures and monitoring and alarm systems. However, all the interactions between component failures tend to be stronger when components are highly loaded and components have smaller margins so they can tolerate only smaller increases in load before failure, the system nonlinearities and dynamical couplings increase, and the system operators have fewer options and more stress.

A typical large blackout has an initial disturbance or trigger event followed by a sequence of cascading events. Each event further weakens and stresses the system and makes subsequent events more likely. The blackout events and interactions are often rare, unusual, or unanticipated because the likely and anticipated failures are already routinely accounted for in power system design and operation. The complexity is such that it can take months after a large blackout to sift through the records, establish the events occurring and reproduce with computer simulations and hindsights a causal sequence of events: components that fail when their load exceeds a threshold, an initial disturbance loading the system, and the additional loading of components by the failure of other components. The blackout cascades in this model are essentially instantaneous events due to dynamical redistribution of power flows and are triggered by probabilistic failures of overloaded lines. Inevitably, some of the salient features of cascading failure in blackouts need to be treated with probabilistic models (Dobson et al. 2005).

6.4.3.1 Models for Describing Failure Cascade Dynamics

An example of a basic cascading failure model, proposed by Dobson et al. (2005), considers a system of N identical components with random initial loads sampled uniformly between a minimum value $L_{\min}$ and a maximum value $L_{\max}$. All components have the same limit of operation L_{fail}, beyond which they are failed. To start the cascade, an initial disturbance imposes on each component an additional load D. If the sum of the initial load L_j of component j and the disturbance D is larger than a component load threshold L_{fail}, component j fails. When a component fails, a fixed and positive amount of load P is transferred to each of the system components.

In an extension of the model, proposed by Zio and Sansavini (2008), the overload P is propagated locally to first-neighbors of the failed node in the network structure. If there is no working node in the neighborhood of a failed component, the cascade spreading in that "direction" is stopped. The transfer of a fixed amount of load P to neighboring components of the failed one may be representative of what happens in systems where each node equally contributes to the global system activity and following their progressive failures the same amount of damage is caused to the still working ones. The case of a fully connected system, where all nodes are first-neighbors, coincides with the original model proposed in Dobson et al. (2005).

The algorithm for simulating the dynamics of cascading failures in the extended model proceeds in successive stages as follows:

(1) At stage $i = 0$, all N components are initially working under independent uniformly random initial loads $L_1, L_2, \ldots, L_N \in [L_{\min}, L_{\max}]$, with $L_{\max} < L_{\text{fail}}$.
(2) An initial disturbance D is added to the load of each component.
(3) Each unfailed component is tested for failure: for $j = 1, \ldots, N$, if component j is unfailed and its load $> L_{\text{fail}}$ then component j fails.
(4) The components loads are incremented taking into account the network topology, i.e. the failed component neighborhood: for each failed node, the load of its first-neighbors is incremented by an amount P. If the neighborhood set of the failed node is empty, the associated failure propagation comes to an end.

For the probabilistic dynamics modeling of the failure cascade process, the above cascade propagation algorithm is embedded in a Monte Carlo simulation framework, in which a large number of cascade processes are triggered for the same range of initial load, $[L_{\min}, L_{\max}]$, in order to obtain statistically significant results. The damage caused by the cascades for any initial load level is quantified in terms of the number of network components which have failed on the average, i.e. the average cascade size, $\bar{S}$. The simulation is repeated for different ranges of initial load, with $L_{\max} = 1$ and $L_{\min}$ varying from 0 to 1, and a point $(L, \bar{S})$ is drawn in a load-size diagram, so that the average critical load, L_{cr}, at which the phase transition between the absence of cascades and the emergence of cascades with significant size ($\bar{S} \geq S_{cr}$, e.g. involving a relevant fraction of network components) in the system can be identified.

A further modification of the model is proposed in Zio and Sansavini (2008) to consider systems for which it is more realistic that the actual load (and not a fixed amount P) previously carried by the failed component is passed onto the neighboring components in the network. To model such condition, step 3 of the cascade propagation algorithm is modified as follows:

(3) The components loads are incremented taking into account the network topology: given the generic node j, failed under load $L_j^* > L_{\text{fail}}$, its load L_j^* is spread uniformly among its first neighbors, by incrementing their load of an amount equal to L_j^* divided by the degree k_j of the failed node. If the neighborhood set of the failed node is empty, the associated failure propagation comes to an end.

In other words, the load of the failed component is uniformly shared among its neighbors. It still holds that in case of an empty neighborhood, the load is no longer propagated and the cascade is stopped in that "direction".

The probabilistic dynamics modeling of failure cascades, framed within a scheme of computational Monte Carlo simulation, allows delving into the failure cascade unfolding and devising indicators of component criticality for the cascade process. Four such indicators are illustrated below (Zio and Sansavini 2011a).

The frequency of participation to a cascade, f_i, of every component $i = 1, 2, \ldots, N$ can be evaluated normalizing the number of its failures over the number of failure cascades simulated starting from different initial conditions (load disturbance or component attacked, depending on the model):

$$f_i = \frac{\text{number of failures of component i}}{\text{number of cascades simulated}} \tag{6.21}$$

This measure gives information about the importance of a component in the buildup of a cascade.

The average time, t_i, of entrance into a cascade of a component $i = 1, 2, \ldots, N$ can be assessed averaging over the total number of cascades simulated:

$$t_i = \frac{\text{time when component i enters the cascade}}{\text{total number of cascades simulated}} \tag{6.22}$$

This indicator is a measure of how early in time a component gets involved in a cascade process.

To catch how the failure of the generic component $i = 1, 2, \ldots, N$ causes other components to fail subsequently, the average duration, d_i, and final size, s_i, of a cascade following the failure of component i can be evaluated through the same averaging procedure used for computing the average time of entrance:

$$d_i = \frac{\text{duration of a cascade following the failure of component i}}{\text{total number of cascades simulated}} \tag{6.23}$$

$$s_i = \frac{\text{final size of a cascade following the failure of component i}}{\text{total number of cascades simulated}} \tag{6.24}$$

It is expected that the failures of more critical components will result in larger sizes of the developing cascades; furthermore, two different behaviors can be anticipated for the duration of the emergent cascade: namely, the failure of a critical component could lead either to the sudden failure of the remaining working components, with a very short cascade duration or to a long chain of delayed failures, resulting in a long cascade duration. In this sense, the final size of the cascade is considered a direct indicator of the criticality of a component whereas the duration measure by itself does not allow drawing clear-cut conclusions about the critical contribution of components to the cascading failure process.

It is important to stress once more that the two averages in the indicators Eqs. 6.23 and 6.24 are taken with respect to the total number of cascades triggered in the system, to reflect the component average relevance to the cascade process.

Evaluation of these indicators for some case studies of literature have shown that when applied to models of local propagation of a fixed amount of load and of redistribution of the failure load, three of the proposed criticality indicators, namely f_i, d_i and s_i, are consistent among themselves in their criticality ranking of the components; when applied to the model of cascading failures due to targeted intentional attacks s_i and d_i are consistent in ranking the components according to their criticality (see Sect. 6.4.4 for an example).

In general, the frequency of participation of every component in a cascade, f_i, seems to be the most relevant indicator since it highlights the direct contributions of each component to the cascading failure process irrespective of the different propagation logic. In the case studies considered of random failure propagations, f_i identifies as most critical those nodes with highest degree, while for the intentional attacks to system nodes, it ranks as most critical those which have few connections but which are linked to highly connected components or lie along a direct path linking highly connected components.

Complementary criticality information is provided by the d_i and s_i indicators, which capture the components failure contribution in promoting successive failures. In a reference case study considered in Zio and Sansavini (2011a), for the model of local propagation of a fixed amount of load, s_i particularly highlights those nodes bridging different loosely connected subsets of components, while for the situation of intentional attacks, s_i identifies the criticality of those nodes connected to nodes having high centrality values. Conversely, the ranking provided by the t_i indicator is dependent on the coupling strength among components: when the components are weakly coupled, it gives consistent results with the other indicators, whereas it gives opposite results if the components are strongly coupled. In this respect, the t_i indicator could be useful in identifying the degree of coupling among components in interconnected systems with respect to propagating failures.

6.4.3.2 Critical Loading

As load increases, a cascading failure becomes more likely, not necessarily gradually and/or uniformly but at a point of criticality or phase transition. In complex systems and statistical physics, this point is associated with power tails in probability distributions.

The importance of the critical loading is that it defines a reference point for increasing risk of cascading failure. The terminology of "criticality" and the term "phase transitions" comes from statistical physics and they are used in the sense of "an abrupt change in a global system property": they should be interpreted as a sharp threshold rather than the definition used in statistical mechanics (Huang 1987).

The presence of phase transitions and the presence and intensity of the phase transitions are strongly dependent on system parameters. For example, Internet can be considered as a large collection of autonomous systems (ASs) held together by routing infrastructures. The Internet routing protocols maintain connectivity between and within ASs, and are designed to automatically reconfigure and re-compute routing tables when they detect a link failure. This computation starts locally around the failure point and then the information propagates through the Internet. Physical connectivity failure (link failure, router crash), transient connectivity problems due to congestion or even manual reboots, etc., may result in the delay of message to the peers. A finite set of N routers, all connected to each other, is said to form a clique. It has been observed that the propensity for phase transitions increases as clique size increases and also as the processing capacity of the routers decreases (Coffman et al. 2002).

The models of cascading dynamics indicate that both the size of the clique N_c as well as the capacity of the nodes in the clique is an important consideration for the phase transitions. The size of the clique acts as a threshold for phase transition, given other parameters: the clique must be large enough for the transition to appear. Increasing the clique sizes beyond the threshold does not change the location of the phase transition, but does have an effect on relative stability. On the other hand, if the clique size is large enough, then the capacity of the nodes in the system decides the location where the phase transition occurs.

6.4.3.3 Self-Organization and Mitigation

The probability of component failure in power systems generally increases with component loading. Each failure is a limiting or zeroing of load in a component and causes a redistribution of power flow in the network and hence an increase in the loading of other system components. If the cascade of events leads to limiting or zeroing the load at substations, it is a blackout.

A cascade of events leading to blackout usually occurs on a time scale of minutes to hours. Efforts to avoid blackouts and especially to avoid repeated blackouts with similar causes include repair of damaged equipment, more frequent maintenance, changes in operating policy away from the specific conditions causing the blackout, installing new equipment to increase system capacity, and adjusting or adding system alarms or controls. These engineering responses to a blackout occur on a range of time scales longer than one day. They reduce the probability of events in components related to the blackout, either by lowering their probabilities directly or by reducing component loading by increasing component capacity or by transferring some of the loading to other components. Thus, the probability of a similar blackout occurring is reduced, at least until load growth degrades the improvements made.

The pattern or vector of component loadings may be thought of as a system state. Maximum component loadings are driven up by the slow increase in customer loads via the operating policy. The loadings of components involved in the blackout are reduced or relaxed by the above engineering. However, the loadings of some components not involved in the blackout may increase. These opposing forces driving the component loadings up and relaxing the component loadings are a reflection of the standard tradeoff between satisfying customer loads economically and security and apply over a range of time scales. Dobson et al. (2004) suggest that the opposing forces, together with the underlying growth in customer load and diversity give rise to a dynamic equilibrium.

These ideas of complex dynamics by which the network evolves are inspired by corresponding concepts of self-organized criticality (SOC) in statistical physics (Huang 1987). A self-organized critical system is one in which the nonlinear dynamics in the presence of perturbations organize the overall average system state near to, but not at, the state that is marginal to major disruptions. In self-organized critical systems, the probability of occurrence of large disruptive events decreases as a power function of the event size.

The idea is that the slow, opposing forces of load increase and network upgrade in response to blackouts shape the system operating margins so that cascading blackout sizes occur with a frequency governed approximately by a power law relationship size; that is, these forces drive the system to a dynamic equilibrium just below and near criticality (Dobson et al. 2004). Complex dynamics of failure cascades have important implications for power system control and operation and for the efforts to reduce the risk of blackouts (Dobson et al. 2004).

The success of mitigation efforts in self-organized critical systems is strongly influenced by the dynamics of the system. Unless the mitigation efforts alter the self-organization forces driving the system, the system will be pushed to criticality. To alter those forces with mitigation efforts may be quite difficult: the mitigation efforts can move the system to a new dynamic equilibrium while remaining near criticality and preserving the power tails. Thus, while the absolute frequency of disruptions of all sizes may be reduced, the underlying forces can still cause the relative frequency of large disruptions to small disruptions to remain the same.

Indeed, apparently sensible efforts to reduce the risk of smaller blackouts can sometimes increase the risk of large blackouts. This occurs because the large and small blackouts are not independent but are strongly coupled by the dynamics. The possibility of an overall adverse effect on risk from apparently sensible mitigation efforts shows the importance of accounting for complex system dynamics when devising mitigation schemes.

6.4.4 Exemplary Application to Failure Cascade Dynamics Modeling for a Single CI

As an example of application Zio and Sansavini (2011a), the indicators of component criticality introduced have been computed for the topological network of the 380 kV Italian power transmission network (see Fig. 6.8), considering cascades evolving according to the extended failure propagation models of Sect. 6.4.3.1. This reference network can be modeled by $N = 127$ nodes connected by $K = 342$ links (TERNA 2002; Rosato et al. 2007). In all simulations, the cascading failure evolution has been followed step by step, the relevant information collected and, eventually, the quantities Eqs. 6.21–6.24 have been computed.

Table 6.11, columns 1–4, reports the results for the cascade model relative to the local propagation of a fixed amount of load to the first neighbors of the failed component. Three out of the four criticality indicators, namely, f_i, d_i and s_i identify the most critical components with respect to the different cascade features they measure. Components 64, 68 and 88 turn out to be the most critical with respect to f_i and d_i while according to s_i component 64 and 88 are less important than other components, e.g. 101 (Villanova in Fig. 6.8); this is due to the fact that the latter constitutes a bridge between different loosely connected subsets of components, namely, between the Northern and the Southern Adriatic backbone, and thus functions as a channel for spreading the failure to regions of the system which are far apart.

The ranking agreement among the f_i and d_i indicators is somewhat unexpected since they are related to different cascade features, namely, the frequency of participation and the duration of the cascade.

It is interesting to note that the average time of entrance t_i is shortest for those components which least participate and contribute to the cascade development (e.g., 127, 117, and 116). According to this propagation model, following a failure, a small extra load is given to the neighboring components and, consequently, the cascade never affects the whole system, in particular sparing the less connected

Table 6.11 Summary of the criticality indicators rankings for the two models of cascading failures (only the 24 most critical nodes are reported)

Propagation of a fixed amount of load				Redistribution of the failure load			
f_i	t_i	d_i	s_i	f_i	t_i	d_i	s_i
64	125	64	68	64	64	64	69
68	126	88	24	68	70	35	70
88	124	68	115	35	68	59	87
67	121	35	43	59	35	60	26
79	123	67	7	60	69	88	74
35	52	79	101	88	88	68	1
60	55	60	3	24	79	79	50
75	115	59	2	79	59	14	4
59	3	98	21	43	43	61	117
81	56	75	64	14	60	43	57
98	2	103	88	61	14	21	77
63	7	97	103	28	110	62	116
62	120	43	35	67	21	67	37
103	122	24	110	21	87	63	19
92	24	81	79	27	98	40	72
97	8	63	52	63	67	98	93
91	113	14	28	62	101	110	51
41	68	40	92	103	75	75	54
61	54	101	55	75	97	97	94
14	114	61	120	97	61	24	124
71	127	27	47	98	24	28	44
40	101	28	8	40	40	101	125
43	119	7	125	81	103	27	126
78	21	41	124	101	81	103	121

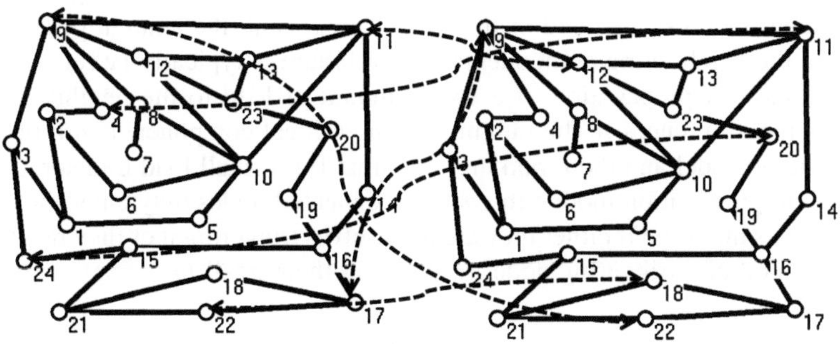

Fig. 6.15 The topological model of the IEEE RTS–96 (Grigg et al. 1996). Each system i has $N_i = 24$ nodes (*circles*) and $K_i = 34$ links (*solid lines*), $i = 1, 2$. The $M = 6$ interdependency links connecting the two systems are also shown (*dashed lines*)

components, e.g., nodes 127, 117, and 116. Thus, the poorly connected nodes only enter the cascade soon after its initiation if either they are themselves triggering it or they reside in the neighborhood of a triggering node: this results in their small average time of entrance in the cascade, t_i (Table 6.11, column 2).

Table 6.11, columns 5–8, report the results for the cascade model which redistributes the failure load onto the neighborhood of the failed node. All the component criticality indicators agree that 64, 68, 35, 59, 60, and 88 are most critical to the cascading process. In particular, nodes 59 (Piacenza) and 60 (Caorso) form a bridge between two densely connected areas in Northern Italy. In this failure propagation model, the amount of load transferred to the neighboring components after a failure is typically larger than in the previous case of a fixed amount of load; this gives rise to a stronger coupling among components so that when a cascade is initiated it is more likely to fully develop and affect the whole system: thus, the components more prone to failure are the ones which are most connected.

It can also be noticed that the nodes which least contribute to the cascade process, nodes 69, 70, 87, and 26 according to f_i, are ranked as having the highest s_i; this is due to the fact that if they get involved in a failure cascade this happens early in time, e.g. nodes 69 and 70 according to t_i, before the cascade has spread over a large portion of the system.

Overall, the ranking results are consistent with a physical analysis of the network system, indeed highlighting the components which most affect the failure spreading. The logics of propagation of a fixed amount of load and of redistribution of the failure load give consistent results across the criticality indicators: in both cases, the most critical components according to f_i, d_i, and s_i are those with highest degree (68, Ravenna Canala and 64, Martignone) and which constitute a bridge between different loosely connected subsets of components, whose failure effect spreads to regions far apart in the system (59–60, 88, Montalto and 79, Poggio a Caiano). Conversely, it is not always true that most connected nodes are the most critical as it can be seen from node 24 (Milano Centro), which is not among the most critical in the fixed-amount-of load redistribution model. In the failure propagation model with redistribution of the load of the failed component, the amount of load transferred to the neighboring components after a failure is typically larger than in the previous case of propagation of a fixed amount of load to the neighboring; this explains the differences in the ranking among critical components with respect to the previous model and the t_i ranking; note that for a small load transfer, as in the first failure propagation model, the poorly connected nodes only enter the cascade soon after its initiation if either they are themselves triggering it or they reside in the neighborhood of a triggering node, resulting in their small t_i.

6.4.5 Probabilistic Dynamic Modeling of Interdependent CIs

Dependencies and interdependencies among different CIs have to be modeled for assessing the influences and limitations which interacting infrastructures impose on the individual system operating conditions, for avoiding fault propagation by designing redundancies and alternative modes of operations, for detecting and recognizing threats (Zimmermann 2001). In developing modeling and simulation frameworks, it is important to know that simply linking existing infrastructure models together fails to capture the emergent behavior arising in interdependent infrastructures, a key element of interdependency analysis. The logical view that the larger coupled system is just a new larger complex system underestimates the heterogeneity introduced through the coupling of the systems. While the individual systems may have a relatively homogeneous structure, the coupling between the systems is often fundamentally different both in terms of spatial uniformity and in terms of coupling strength. Understanding the effect of this coupling on the system dynamics is necessary if we are to accurately develop risk models for the different infrastructure systems individually or collectively (Newman et al. 2005).

Examples of the types of potential coupled infrastructure systems include power–communication systems, power–market systems, communication–transportation systems, and even market–market systems. The effect of the coupling can be critical and obvious for systems that are strongly coupled such as the power–market systems. Perturbations in one can have a rapid and very visible impact on the other. In fact, in many ways such systems are often thought of as one larger system even though the coupling is not homogeneous and each of the component systems (namely, the market and the power transmission systems) can have its own separate perturbations and dynamics. For other less tightly coupled systems, such as power–communication systems, the effect can be much more subtle but still very important. In such systems small perturbations in one might have very little obvious effect on the other system, yet the effect of the coupling of the two systems can have a profound effect on the risk of large, rare disturbances (Newman et al. 2005).

In order to characterize the extent to which a contingency affecting an infrastructure is going to weaken, and possibly disrupt, the safe operation of an interconnected system, it is necessary to model the relations established through the connections linking the multiple components of the involved infrastructures. The modeling of interdependencies among network systems and of their effects on failure propagation can be carried out within the simulation framework of failure cascade processes; the sensitivity of the coupling parameters defining the interdependency strength is of particular interest for the definition and prescription of cascade-safe operating margins in interdependent CIs.

6.4.5.1 Failure Cascade Modeling

In analogy to the case of individual CIs, the dynamic modeling of interdependent CIs can be carried out in a cascading failure simulation framework which abstracts the physical details of the services provided by the individual infrastructures, while at the same time capturing their essential operating features and interdependencies, and examines the emergent effects of disruptions, with the associated downstream consequences (Newman et al. 2005; Zio and Sansavini 2011b).

In such framework, interdependencies are modeled as links connecting nodes of the interdependent systems; these links are conceptually similar to those of the individual systems and can be bidirectional with respect to the "flow" between the interdependent networks. Cascading failures are then assessed considering the local propagation of the overload originated from a failure to first-neighbors and to the interdependent set of components linked to the failed one.

The number of interdependency links, *M*, and the load of flow transferred over the interdependency links, *I*, are essential features characterizing the "coupling energy" between the interdependent systems.

The interdependencies analysis can be carried out among as many CIs as needed; as illustrative example we shall consider for simplicity two interdependent systems with fixed "interdependency energy", i.e. such that each node in system 1 could be interdependent with any other node in system 2, but the number of available interdependency links, *M*, between the systems is fixed. Communication systems, in which each agent in system 1 can interact with any other agent in system 2 but there is a maximum amount of connecting energy between the two systems, could be an example of such energy-limited systems.

The interdependent CIs analysis here illustrated focuses on cascade onset and propagation over the topological structures of two interdependent network systems of N_1 and N_2 identical components connected by K_1 and K_2 links, with random initial loads sampled uniformly between a minimum value $L^i_{\min}$ and a maximum value $L^i_{\max}, i = 1,\ 2$. No reference is made to the specific flow which characterizes the infrastructures (e.g. electrical current in power transmission networks).

All components in the *i*th system are assumed to have the same limit of operation $L^i{}_{fail}$, beyond which they failed $L^1{}_{\text{fail}} = L^2{}_{\text{fail}} = L_{\text{fail}} = 1$ in this example, upon normalization of all loads relative to the failure load value). Let us also assume that the overload is propagated locally, to the first-neighbors of the failed node within the network structure it belongs to (a fixed and positive amount of load, P^i, $i = 1,\ 2$) and to the interdependent components which the failed component is connected to in the other network system (a fixed and positive amount of load, *I*), if any. If there is no working node in the neighborhood of a failed component or among the interdependent nodes connected to it, the cascade spreading along that path is stopped. The case of two fully connected systems, where all nodes are first-neighbors and every component in system 1 is interdependent to every component in system 2, coincides with the model proposed in Newman et al. (2005).

The interdependency links between the two network systems are treated in the same way as the individual system links. They are bidirectional connections and upon the failure of a node in system 1 or 2, the overload, *I*, is propagated locally to the nodes in the interdependent network system 2 or 1, if any interdependency is present for the failed node. This transfer accomplishes the coupling between the two systems.

To account for the dynamics of changing connections between the two systems under developing failure cascade processes, Monte Carlo simulations can be performed in which the number of interdependency links, M, and the load transferred over the interdependency links, I, are kept constant but the interdependency connections among components are randomly rewired at each trial.

To start the cascade, an initial disturbance imposes an additional load $D^i_{j_i}$ on each component j_i of the two systems, $j_i = 1, 2, \ldots, N_i$, $i = 1, 2$. If the sum of the initial load $L^i_{j_i}$ and the disturbance $D^i_{j_i}$ of component j_i in system $i = 1, 2$ is larger than the component load threshold L^i_{fail}, component j_i fails. This failure occurrence leads to the redistribution of additional loads, P^i on the neighboring nodes and I on the interdependent nodes, which may, in turn, get overloaded and thus fail in a cascade which follows the connection and interdependency patterns of the network systems. As the components become progressively more loaded, the cascade proceeds.

The algorithm for simulating the cascading failures proceeds in successive stages as follows:

(0) At stage $m = 0$, all $N_1 + N_2$ components in systems 1 and 2 are working under independent uniformly random initial loads $L^i_1, L^i_2, \ldots, L^i_{N_i}\left[L^i_{\min}, L^i_{\max}\right]$, with $L^i_{\max} < L^i_{\text{fail}}, i = 1, 2$.
(1) M interdependency links between system 1 and system 2 are generated, connecting two randomly selected components.
(2) An initial disturbance D^i, $i = 1, 2$, is added to the load of each component in the two systems.
(3) Each unfailed component is tested for failure: for $j_i = 1, \ldots, N_i$, if component j_i is unfailed and its load $L^i_{ji} > L^i_{\text{fail}}$, then component j_i in system i fails, $i = 1, 2$.
(4) The components loads are incremented taking into account the network topology: for each failed node in system i, the load of its first-neighbors is incremented by an amount P^i, $i = 1, 2$. If the working neighborhood set of the failed node is empty, the associated failure propagation into the system comes to an end.
(5) The components loads are incremented taking into account the interdependency pattern: for each failed node in system 1 or 2, the load of its interdependent nodes in system 2 or 1 is incremented by an amount I. If the interdependency set of the working components of the failed node is empty, the associated failure propagation to the interdependent system comes to an end.
(6) The stage counter m is incremented by 1 and the algorithm is returned to step 3.

The algorithm stops when failures are no further propagated intra or inter the two systems.

Various initial system loading levels can be evaluated varying the uniform sampling ranges $\left[L^i_{\min}, L^i_{\max}\right], i = 1,\ 2$, whose midpoints, L^i, are indicators of the average initial systems loading levels. Large L^i values relate to operating conditions in which the systems are more stressed.

The effects of the interdependencies between the two systems can be analyzed in terms of the average cascade size, S^i, i.e., the number of failed components in the ith system at the end of the cascade spread, versus the average initial load in the system, L^i, which represents the system operating level, $i = 1, 2$. For each value of L^i, several Monte Carlo simulations must be repeated, each simulation corresponding to a different sampled pattern of the M interdependency links.

It is expected that the interdependencies cause a shift to lower values of the loading threshold for which the cascading phenomenon starts appearing with significance. To quantitatively assess the effects of the interdependency, a threshold representing the maximum allowable cascade size, $S^i{}_{cr}$, can be set which identifies the critical load, $L^i{}_{cr}$, beyond which the threshold is exceeded in system $i = 1, 2$. The maximum allowable cascade size, $S^i{}_{cr}$, is interpreted as the maximum number of components which can be lost in system i without affecting the global service provided by the infrastructure. This threshold can vary from system to system and is a distinguishing feature of the provided service.

The critical load, L^i_{cr}, is a relevant feature of a network system since it identifies, together with the continuous change in gradient, a type-two transition (Huang 1987) between the cascade-safe region and the onset of disrupting cascades in terms of the loading conditions, L^i. Along with the average cascade size, S^i, it gives essential information on the system vulnerability toward cascading failures and it can help identifying safety margins of system operation.

To understand the effects on the cascade process of the parameters characterizing the interdependency between the two systems, a further sensitivity analysis can be performed in several operating conditions which reflect real system operations.

Firstly, it is important to assess the extent to which an interdependent system working at different, fixed load levels influences the coupled network system with respect to its vulnerability toward cascading failures. To this aim, the variation of the critical load, L^1_{cr}, of system 1 can be assessed while system 2 is working at fixed constant loads. It is important to understand that the coupling between the two interdependent systems is such that under given loading conditions, beyond certain thresholds, the effects of the systems nonlinearities become relevant and an emergent behavior arises in the interdependent systems. When designing and operating interdependent infrastructures, it is then necessary to control the operating levels of the systems and assess the values beyond which nonlinearities start governing the system cascading failure behavior.

Then, the effects of the number of interdependency links, M, on the vulnerability to cascading failures can be assessed in two different system operating conditions. The characterization of this relationship is relevant in the definition of cascade-safe operating regimes for the interdependent systems: for a fixed number of interdependency links in the system, M, a critical loading level can be identified below which the systems are safely operated. Cascade-safe operating regimes can be identified also with respect to the average cascade size, S; once the operating level is known the systems can be operated or designed to limit the maximum average cascade size, S.

The effects of the load transferred upon failure over the interdependency links I, of the failed component can also be assessed, even if the load transferred over the interdependency links, is a less critical parameter in designing and operating

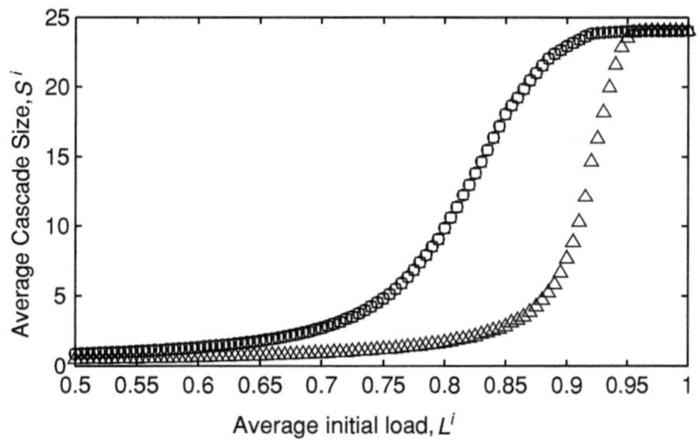

Fig. 6.16 The average cascade size, S^i versus the average initial load in the system, L^i, $i = 1, 2$. *Triangles* isolated single system. *Squares* and *circles* identical interdependent systems

interdependent network systems against cascading failures (Zio and Sansavini 2011b). As before, cascade-safe operating regimes for the systems can be identified with respect to the critical load, L_{cr}: for a fixed value of the load transferred over the interdependency links, I, a critical loading level can be determined below which the systems can be safely operated. Also with respect to system parameter I, cascade-safe operating regimes can be identified with respect to the average cascade size, S: given the operating level, the systems can be operated or designed to limit the maximum S.

6.4.5.2 Exemplary Application to Two Interdependent CIs

The model of cascading failures in interdependent network systems is applied on a modified literature case study with the aim of identifying the interdependency features most critical for the cascade-safe operations of interdependent infrastructures and defining related cascade-safe operational margins (Zio and Sansavini 2011b). In Fig. 6.15, the systems considered are two identical networks which are an abstract topological model of the IEEE Reliability Test System–96 (RTS-96) (Grigg et al. 1996). M interdependency links are drawn between them as explained above (*dashed lines* in Fig. 6.15). In the proposed analysis, interest is on cascade onset and propagation over the bare topological structure of the test systems; no reference is made to the specific electrical flow which characterizes these electrical infrastructures.

Consider two systems of N_1 and N_2 identical components ($N_1 = N_2 = N = 24$ in this study) connected by K_1 and K_2 identical links ($K_1 = K_2 = K = 34$) with random initial loads sampled uniformly between a minimum value $L^i_{\min}$ and a maximum value $L^i_{\max}$. The two systems are connected by M interdependency links ($M = 34$ in the study, except during the sensitivity analysis with respect to changing M). The initial disturbance imposes an additional load $D^i_{j_i}$ on each component j_i of the two systems, $j_i = 1, 2, \ldots, N_i$, $i = 1, 2$ ($D^1_{j_1} = D^2_{j_2}$ $D = 0.02$ in this study). Moreover, when the systems are operating at varying average initial load, L^i, its range of variation is [0.5, 1] at steps of 0.005.

Fig. 6.17 Critical load, L^1_{cr}, in system 1 for constant average initial load levels, L^2

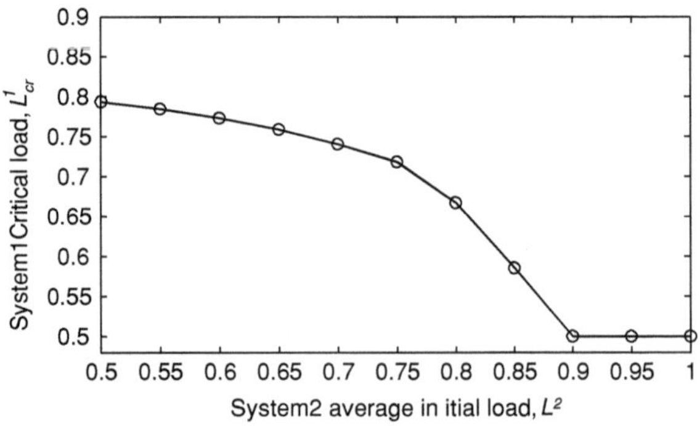

The effects of the interdependencies between the two systems are shown in Fig. 6.16 in terms of the average cascade size, S^i, i.e., the number of failed components in the ith system at the end of the cascade spread, versus the average initial load in the system, L^i, which represents the system operating level, $i = 1, 2$. For each value of L^i, varying in the range [0.5, 1] at steps of 0.005, 100,000 Monte Carlo simulations have been repeated, each simulation corresponding to a different sampled pattern of the M interdependency links.

The triangles represent the average cascade size, S^i, in system $i = 1, 2$ as a function of the average initial load, $L^i = L$, for the isolated single system $i = 1, 2$, i.e. when no interdependency is present. The overlapping squares and circles represent the same quantity $S^1 = S^2$ for the identical and identically operating systems 1 and 2, respectively.

As expected, the interdependencies cause a shift to lower values of the loading threshold for which the cascading phenomenon starts appearing with significance (from approximately 0.9 for the individual isolated system to approximately 0.8 for the interdependent systems). Notice that as the average initial loading, $L^i = L$, on the system increases (with a smoother behavior for the two interdependent networks than for the individual isolated system, due to the fact that cascades start arising at lower average initial loading, $L^i = L$, in the interdependent networks which, thus, are less stressed and prone to their propagation), the systems are increasingly vulnerable to cascading failures.

To quantitatively assess the effects of the interdependency in Fig. 6.16, a threshold representing the maximum allowable cascade size, S^i_{cr}, can be set which identifies the critical load, L^i_{cr}, beyond which the threshold is exceeded in system $i = 1, 2$. The maximum allowable cascade size, S^i_{cr}, is interpreted as the maximum number of components which can be lost in system i without affecting the global service provided by the infrastructure. This threshold can vary from system to system and is a distinguishing feature of the provided service. In the following, for simplicity but with no loss of generality $S^1{}_{cr} = S^2{}_{cr} = S_{cr} = 15\%$ is assumed which identifies $L^1{}_{cr} = L^2{}_{cr} = L_{cr} = 0.8662$ for the individual systems in isolated conditions and $L^1{}_{cr} = L^2{}_{cr} = L_{cr} = 0.7266$, for the two interdependent systems. As previously explained in Sect. 6.4.3.1 and Sect. 6.4.3.2 the critical load, L^i_{cr}, is a relevant

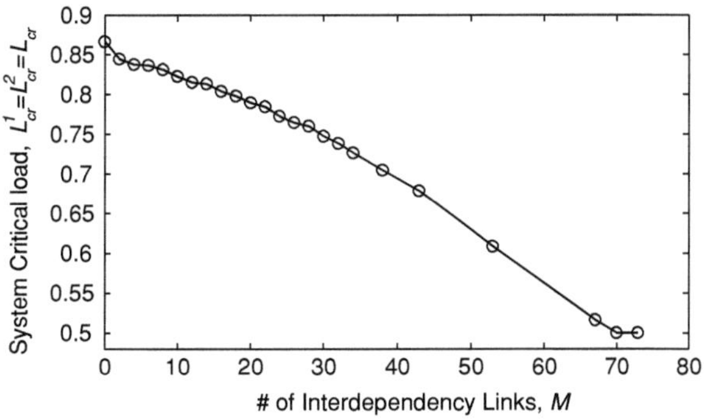

Fig. 6.18 Critical load, $L^1_{cr} = L^2_{cr} = L_{cr}$, in systems 1 and 2 versus the number of interdependency links, M. Systems are working at the same varying average initial loads, $L^1 = L^2 = L$ (load transferred over interdependency links, I, equals 0.07)

feature of a network system since it identifies, together with the continuous change in gradient, a type-two transition between the cascade-safe region and the onset of disrupting cascades in terms of the loading conditions, L^i. Along with the average cascade size, S^i, it gives essential information on the system vulnerability toward cascading failures and it can help identifying safety margins of system operation.

To understand the effects on the cascade process of the parameters characterizing the interdependency between the two systems, further sensitivity analyses can be performed for conditions which reflect real system operations. A first analysis aims at assessing the extent to which an interdependent system working at different, fixed load levels influences the coupled network system with respect to its vulnerability toward cascading failures. To this aim, the variation of the critical load, $L^1{}_{cr}$, of system 1 is assessed while system 2 is working at fixed constant loads; the analysis is performed crudely for fixed values of average initial loads of system 2 ranging between $L^2 = 0.5$ and $L^2 = 1$, in steps of 0.05.

In Fig. 6.17, the results of this analysis are shown. As expected, the coupling between the two systems weakens the resistance of system 1 to failure cascade, forcing it to be operated at increasingly lower levels as the average initial load of system 2 increases. The emerging functional dependence, however, could not be easily anticipated, with a smooth, linear decrease in $L^1{}_{cr}$ for system 2 loading levels below $L^2 = 0.75$, changing to a sudden drop in $L^1{}_{cr}$ when system 2 loading levels rise above $L^2 = 0.75$ followed by the saturation in system 1 $L^1{}_{cr}$ value for system 2 average initial loads beyond $L^2 = 0.9$, indicating that under these conditions of loading on system 2, system 1 experiences unbearable cascades irrespective of its loading level, i.e. there is no cascade-safe region for system 1 when system 2 is operating beyond 90% of the component failure load, $L^2{}_{fail}$.

The trend found indicates that the coupling between the two interdependent systems is such that under given loading conditions, beyond certain thresholds, the effects of the systems nonlinearities become relevant and an emergent behavior arises in the interdependent systems.

A second sensitivity analysis can look into the effects of the number of interdependency links, M, on the vulnerability to cascading failures, in two different system operating conditions.

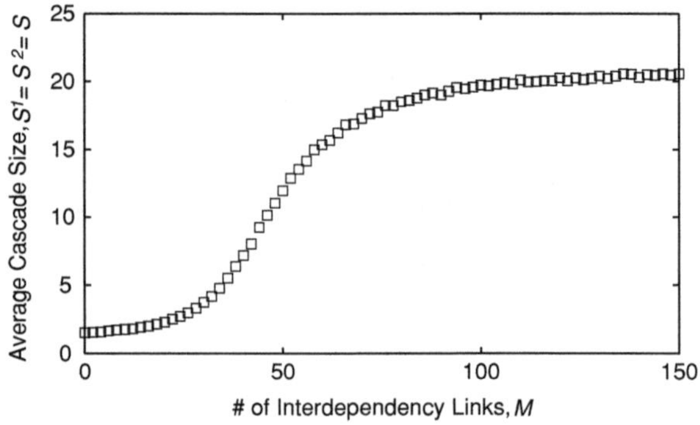

Fig. 6.19 Average cascade size, $S^1 = S^2 = S$, in systems 1 and 2 versus the number of interdependency links, M, while systems operate at the same constant working load, $L^1 = L^2 = L = 0.75$ (load transferred over interdependency links, I, equals 0.07)

In the first case, both systems are operating at the same varying average initial load, $L^1 = L^2 = L$ [0.5, 1] at steps of 0.005 and the variation of system 1 critical load, $L^1{}_{\text{cr}}$, is analyzed with respect to the number of interdependency links, M (Fig. 6.18). Since both systems are identical and operate at the same loading conditions, they will show identical trends of critical load, $L^1{}_{\text{cr}} = L^2{}_{\text{cr}} = L_{\text{cr}}$, similarly to the behavior of Fig. 6.16. It can be seen from Fig. 6.18 that there is an approximately linear functional relationship between the systems critical load, L_{cr}, and the number of interdependency links, M, up to the value $M = 70$ for which the systems cascade-safe region disappears, meaning that the systems are going to experience unbearable cascades irrespective of the loading level, i.e. there is no cascade-safe region for the systems when more than $M = 70$ interdependency links are present between the two. Thus, if one were to try to protect the interdependent systems from cascade failure by controlling the number of their interdependency links M, it appears that nonlinearities do not play a significant role and a linear decrease of the cascade-safe region is to be characterized with respect to the addition of interdependency links between the two systems.

The characterization of this relationship is relevant in the definition of cascade-safe operating regimes for the interdependent systems: for a fixed number of interdependency links in the system, M, a critical loading level can be identified below which the systems can be safely operated. In the present example, it turns out that there is no safety margin when more than $M = 70$ interdependency links are present between the two systems, which is more than twice the number of links in each system, $K_i = 34$, $i = 1, 2$.

In the second case, both systems are operating at the same constant average initial load $L^1 = L^2 = L = 75\%$ of the component failure load, $L^1_{\text{fail}} = L^2_{\text{fail}} = L_{\text{fail}}$, and the average cascade size, $S^1 = S^2 = S$, is assessed with respect to the variation of the number of interdependency links, M (Fig. 6.19). As expected, the average cascade size, $S^1 = S^2 = S$, increases as the number of interdependency links, M, increases until a saturation value is reached which is a function of the load transferred over the interdependency links, I, and the systems constant average initial load, L.

Cascade-safe operating regimes can be identified with respect to the average cascade size, S; once the operating level is known ($L = 75\%$ in this case), the systems can be operated or designed to limit the maximum average cascade size, S. As an example, from Fig. 6.19 it can be understood that in order to have cascades involving less than 15% of the system components, no more than $M = 31$ interdependency links can be operated between the two systems.

6.4.6 Conclusions

Probabilistic dynamics modeling provides a comprehensive representation of the failure behavior in a CI as it emerges from an initial perturbation cascading failures.

Various parameters influencing the failure cascade dynamics can be analyzed. In particular, the probabilistic dynamics model can identify the critical loadings at which the probability of cascading failure increases: determining the proximity to critical loading from power system simulations or data is an important issue for network control and stability. The differences in the propagation behavior of the failure load greatly affect the outbreak and the size of the emerging cascade: knowing this feature is worthy of increase efficiency in both preventing and mitigating cascading failures.

Special attention should also be placed on interdependencies within or among systems and on how coupling between the systems modifies the conditions of safe operation. CIs are characterized by critical loadings and they must operate well-below these values to avoid "normal accidents" and large scale failures; coupling between systems and the effect of the heterogeneity introduced through the different properties of each individual system change the values of these critical loadings.

6.5 Agent-Based Modeling and Simulation

6.5.1 Conceptual Outline

Agent-based modeling (ABM) is a powerful technique for the computerized simulation of large-scale complex systems. It aims to replicate the behavior of real-world systems by modeling its components as a collection of autonomous entities, called agents, as well as the interactions among them. Thus, depending on the nature of the specific system and the goal of the analysis, agents can be as diverse as needed. Examples range from individuals to whole organizations and from single technical components to larger subsystems. The overall system behavior emerges from the simulated interactions of the individual agents, based on the attributes and rules assigned to them.

The roots of ABM can be traced back to the 1940s when cellular automata where used to simulate grids of two-states switches interacting with their nearest neighbors (Buchanan 2009). It was eventually in the 1990s that the boost of computational power paved the way for massive application of these models, particularly in areas like social science, biology and economics. Just recently,

ABM has also been advocated to constitute an indispensable technique for capturing the intricate behavior of socio-technical systems (Kröger 2008), in which different technological layers coexist with their social components that drive their use and development (Vespignani 2009).

ABM is particularly useful when: (Bonabeau 2002)

- Individual behavior is governed by nonlinearities and can be described by thresholds, if–then rules, or nonlinear coupling. Describing such discontinuities is difficult or even impossible with differential equations.
- The large number of system elements with highly diverse behaviour prohibits a tractable description of the system with differential equations.
- The overall system behavior is rooted in its underlying network topology as well as in the nature of the dynamical processes taking place on top of it.
- Individual behavior exhibits memory, temporal correlations, learning and adaptation.
- The global behavior of the system shows or is expected to show emergent phenomena as a result of local component interactions.

Considering these general system characteristics, ABM is advocated to constitute an attractive approach for the vulnerability analysis of CIs while tackling the specific challenges given by their intrinsic complexity (see Chap. 3).

6.5.2 Modeling and Simulation Procedure

6.5.2.1 Conceptual Basics

ABM is rather a way of thinking than a well-defined technique. In contrast to traditional top-down approaches which, for instance, use sets of differential equations to model the overall system dynamics, ABM is strongly bottom-up oriented and describes the system under study from the perspective of its components' activities in a highly natural way (e.g., Garrido 2009).

Hence, the underlying mindset is to represent the components and their individual behavior through agents at the microscopic level, and observe the aggregated system behavior as a result of their interactions at the macroscopic level.

6.5.2.2 Defining an Agent

There is no standardized definition of the term agent in the related literature, whereas the differentiation between the terms "agent" and "object" is often ambiguous.[1]

[1] There is no commonly accepted distinction between these two terms (Kaegi et al. 2009). However, according to D'Inverno and Luck (2004), an object can be regarded as an agent without goals.

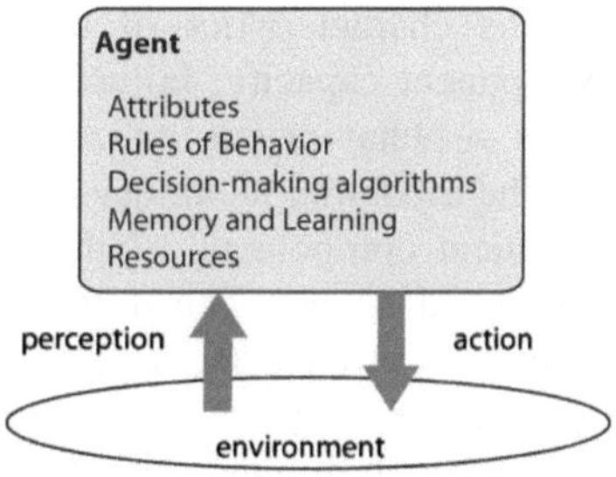

Fig. 6.20 Basic agent concept (adapted from Macal and North 2005)

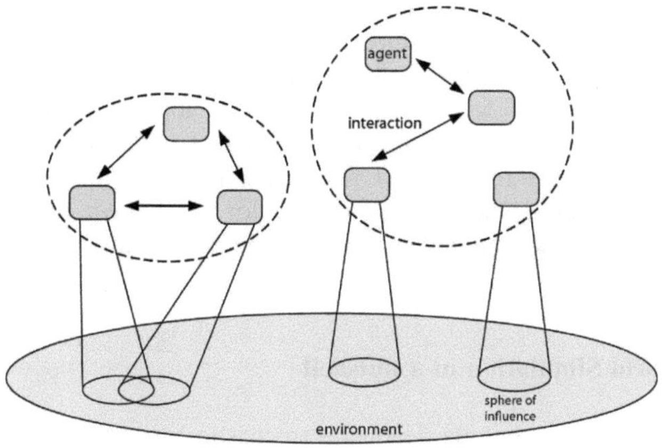

Fig. 6.21 Canonical view of a multi-agent system (adapted from Jennings 2000)

Here, we define an agent as a discrete entity with specific characteristics and a set of rules which govern its behavior and decision-making to achieve its goals. Moreover an agent is autonomous and interacts with other agents within its environment, while being able to learn and adapt its behavior to dynamically changing conditions.

Agents are essentially defined by attributes like geographical location, knowledge, or preferences and rules of behavior like decision-making algorithms or behavioral probabilities. Figure 6.20 depicts the most important characteristics of an agent.

Agents are heterogeneous and thus can model both diverse technical components (e.g., generators in an electric power system) and non-technical components (e.g., transmission system operators in an electric power system).

Figure 6.21 depicts a canonical view of a system comprised of many interacting agents. Those agents that interact with each other form an organizational relationship. Furthermore, the different agents act in an environment with distinct spheres of influence. Based on their interactions, the spheres of influence may intersect or not.

The characteristics of agents are implemented by attributes as diverse as component capacity, failure thresholds and knowledge of system operators. The rules of behavior are realized by decision-making algorithms.

The modeling technique consists of identifying the relevant technical and non-technical components of the system under analysis, and of describing them as individual agents. The unified modeling language (UML) (http://www.uml.org/; Cardellini et al. 2007) can be used as the standardized graphical notation of the agent states and behavioral rules. A state describes the current situation of the agent; it determines, for instance, whether a modeled technical component is in maintenance or not. The rules of behavior can be graphically represented by using finite state machines (FSM) and include both deterministic and stochastic time-dependent, discrete events. A deterministic event is, for instance, the outage of a component when reaching a failure threshold, while stochastic processes are probabilistic component failure models which can be simulated by Monte Carlo techniques (see Box 6.1 for a short introduction to Monte Carlo simulation). During the simulation, the different agents interact with each other directly or indirectly. An example for a direct interaction is the generator dispatch in an electric power system, whereas an example for an indirect interaction are the physical laws governing the power flows in electricity grids.

Monte Carlo Simulation in a nutshell

The Monte Carlo technique denotes a stochastic simulation using algorithmically generated random numbers. Below, a simple example for estimating the unavailability Q of a system:

Assume a system consisting of N components, where s_i is the boolean state of the ith component and Q_i its failure probability. The state of the component is then determined by drawing a random number $R_i \sim$ uniform [0,1] so that

$$s_i \begin{cases} 0 & \text{success} \quad \text{if} \quad R_i > Q_i \\ 1 & \text{failed} \quad \text{if} \quad 0 \leq R_i \leq Q_i \end{cases}$$

Then, the basic steps of a Monte Carlo simulation are as follows:

1. Sample the states of all components ("throw the dices") to get the system state s^j

$$s^j = \{s_1, \ldots, s_i, \ldots, s_N\}$$

2. At each overall sample j assess the system to judge whether it is in a failure state or not.

$$x_{s^j} = 0 \quad \text{if the system is in an operational state}$$

$$x_{sj} = 1 \quad \text{if the system is in a failure state}$$

3. Perform k system state samples. The unbiased estimate of the system unavailability then is given by:

$$\overline{Q} = \frac{1}{k}\sum_{j=1}^{k} x_{sj}$$

with variance:

$$V(\overline{Q}) = \frac{1}{k}V(x) = \frac{1}{k(k-1)}\sum_{j=1}^{k}\left(x_{sj} - \overline{Q}\right)^2$$

Further reading: Billinton and Li (1994)

6.5.2.3 Implementation Procedure

For the vulnerability analysis of CI the practical implementation procedure can vary according to the specific knowledge about the system under study, the needs of the analysis, as well as according to the specific software tool used (see Sect. 6.5.7). An exemplary and simplified work flow which partitions the implementation procedure into six subsequent basic steps is provided in the following:

Step 1—component identification: The relevant technical and non-technical components of the system under study are identified by a screening analysis and grasped as individual agents.

Example (urban water supply): The relevant technical components of an urban water supply system include the different reservoirs, pumps and pipelines. A non-technical component to be represented as a distinct agent is the system operator which monitors and controls the overall system performance.

Step 2—state determination: The relevant discrete states of each agent are determined.

Example (urban water supply): The two discrete states of a water pump considered to be relevant are "up" state, meaning that it is operational, and "down" state, meaning that it is out of order, respectively.

Step 3—setting the state transitions: The rules of behavior are represented by using finite state machines (FSM) and include both deterministic and stochastic

time-dependent, discrete events. The alternation between the distinct agent states is determined by introducing state transitions, which can be triggered by an elapse of time, by signals from other agents (see step 4) and so on.

Example (urban water supply): A water pump fails due to a random failure after a certain period of operating time (i.e., elapse of time). As soon as maintenance actions have been accomplished the water pump returns to its operational state (i.e., signals from other agents).

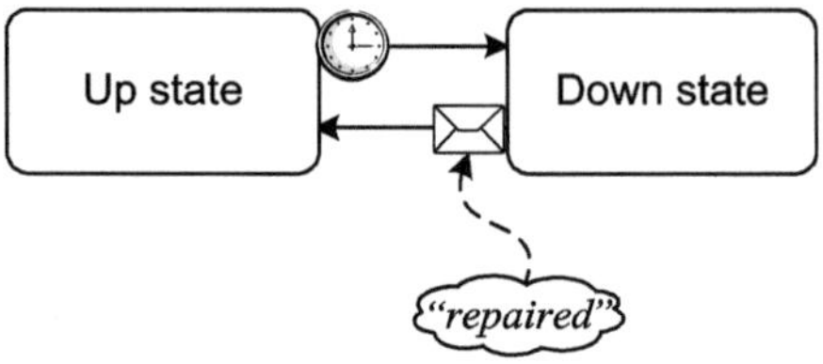

Step 4—establishing agent interactions: Establish the relevant interactions between the agents. Based on real-world system behavior such interactions include message passing (exchange of information) or signal triggering.

Example (urban water supply): Once a water pump fails, the responsible operator becomes informed and subsequently initiates the necessary maintenance action. By the time the pump is repaired, a corresponding message from the operator triggers its transition to the up state.

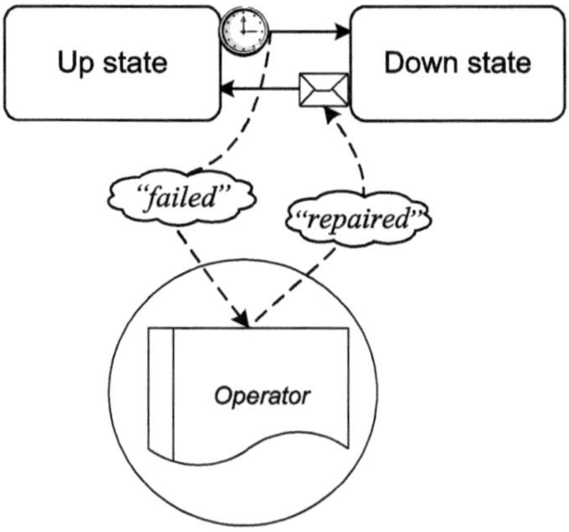

Step 5—running the model: After the simulation model has been developed, the resulting behavior of the agents and their interactions are simulated, generating emerging scenarios which, in turn, may serve as the basis for a holistic vulnerability analysis.

Example (urban water supply): The simulation reveals the complex interplay among the highly diverse components of an urban water supply system responsible for a potential breakdown of the water supply.

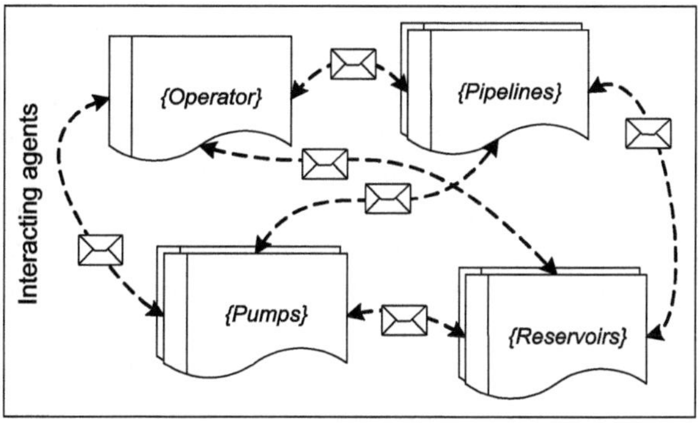

Step 6—benchmarking and calibrating the model: The simulation results are, as far as possible, compared to real-world data. This, in turn, allows for a calibration of the chosen model parameters and further improving the model through a feedback process. Beneath comparison with empirical data gained from the operational use of the system, further validation should be accomplished through expert judgment (see Chap. 5).

Example (urban water supply): Simulated breakdown scenarios are compared to real-world incident statistics. Deviations indicate further model improvements as well as how good it captures the reality.

6.5.3 Simulation Outcome and Expected Results

Given its high level of flexibility, agent-based modeling allows for quantitatively assessing the system under study with respect to a broad spectrum of performance and vulnerability aspects. Examples to be highlighted are as diverse as the identification of critical system components, the estimation of the service unavailability or the emulation of cascading failures within networked systems. This, in turn, allows to assess the impact of different operating strategies of a system while gaining practical insights into those factors with the highest degree of influence on the simulation outcome.

6.5.4 Benefits and Drawbacks

The major benefits of ABM are:

Natural system representation: The model is characterized by a close adherence to the real world and thus can be built "from the bottom up" in a highly natural way. This natural description, in turn, allows experts getting easily acquainted with the built model, being crucial for the further benchmarking and calibration processes.

Tractable analysis of large-scale complex systems: In contrast to describing a system by a set of differential equations whose complexity increases exponentially with the number of possible system states, ABM offers a computationally tractable way of simulating systems that exhibit a large number of dynamic, highly nonlinear interactions among a multitude of heterogeneous components.

Modeling flexibility: ABM features a high flexibility, as the extension or simplification of existing model implementations is straightforward. Moreover, ABM can be iteratively tailored to the aim of a specific study, especially when the full system description is not known ahead of time.

Broad spectrum of possible system studies: An agent-based model of a given system can be readily adapted in such a way that a broad spectrum of different reliability aspects can be assessed, with examples ranging from performance analysis of single components to a multifaceted vulnerability quantification of the whole system under study.

The major drawbacks of ABM are:

Large number of model parameters: The close adherence of the model to the real-world system also implies a large number of different model parameters to be estimated by available empirical data or expert judgment. A plethora of model parameters further complicates the sensitivity analysis of the simulation outputs with respect to their variation.

Operational data requirements: The estimation of the large number of model parameters requires potentially sensitive and confidential operational system data and design information. Therefore, such model data need to be secured by non-disclosure agreements with the respective system operators.

High computational intensity: The explicit consideration of the dynamic interactions among a multitude of system components usually involves long simulation times. The challenge in this respect is to reduce the computational burden by optimizing the technical implementation of the model, e.g., making use of rare event simulation techniques. The evolution of both hardware and software will further fasten up the simulation speed.

6.5.5 State of Development

6.5.5.1 Application to Critical Infrastructures

In the field of CIs, ABM has just recently gained attention for potential applications on a variety of different reliability aspects (e.g., Kröger 2008; Schläpfer et al. 2008). As an example for electric power systems, EMCAS (electricity markets complex adaptive system) uses a large number of agents to model decentralized electricity markets for testing how the behavior of the different agents (e.g., power companies) and the overall system reacts on changing market rules (Argonne National Laboratory 2008). This allows assessing different business models for different markets. Another example is the chemical sector, where ABM has been proposed for capturing the complex interactions inside worldwide networks of chemical production sites, in order to study the systemic performance under a range of business policies and environmental events (Behdani et al. 2010).

Going beyond single critical infrastructures, ABM is regarded to be a promising technique to model and analyze interdependencies among a set of different infrastructure systems (Eusgeld et al. 2009), whereas it may be included into high level architecture (HLA) environments (see Sect. 6.6).

6.5.5.2 Inclusion of Traditional Reliability Methods

Traditional statistical and probabilistic approaches such as fault and event tree methods or reliability block diagrams (Birolini 2007), have long proven to be a suitable approach to quantify the reliability of technical systems. They require to structure the considered failure mechanisms into logic frames (see Sect. 6.3). Such approaches can be useful for determining the stochastic parameters of the behavioral rules of the agents. A vivid example is the use of fault tree analysis for the estimation of the failure probability of a technical component. During the simulation of the multi-agent system, this failure probability then defines the transition time between the up and down states of the component (see step 3 of the modeling and simulation procedure in Sect. 6.5.2).

Apart from serving as static parameter estimation methods, traditional techniques may even be used in a dynamic way by being continuously updated according to the changing agent environment, and providing the ABM with an adjusted feedback information. Consequently, an integrative framework can be established, allowing the inclusion of a broad spectrum of different reliability analysis approaches.

6.5.6 Exemplary Application

This section demonstrates a specific application of ABM for assessing the reliability of an electric power system (EPS) as presented in Schläpfer et al. (2008). Besides further substantiating the benefits and drawbacks of the method, the example might serve as a "role model" for the application to other single-type critical infrastructures or coupled systems.

The main components of EPS being modeled as agents[2] are the generators, loads, transmission lines, busbars and transmission system operators (TSO); as shown by a number of large-area power outages during the last years, the latter play a crucial

[2] In Schläpfer et al. (2008), the term object rather than the term agent is used.

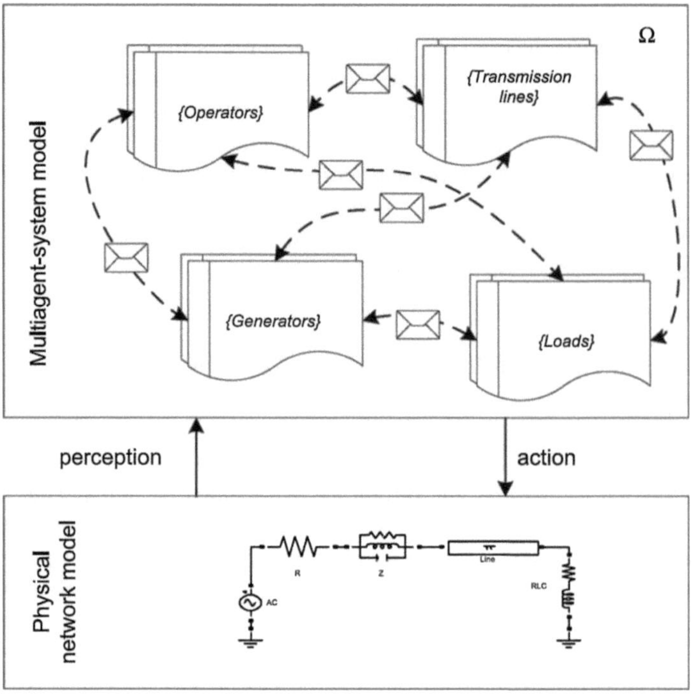

Fig. 6.22 Two-layer-concept applied to the electric power system

role during cascading blackout events and are an example of the importance of non-technical 'components'. In order to combine the agent behavior with the physical laws of the transmission system, a two-layer concept is applied (Fig. 6.22). The lower layer represents the separate modeling of the physical components by means of conventional, deterministic techniques such as power flow calculations whereas the upper layer represents the abstraction of the electric power system with its technical and non-technical components as individual agents. Following the working steps given in Sect. 6.5.2, specific rules of behavior are given to each agent, which include both deterministic and stochastic time-dependent processes, triggered by laps of time or a signal from outside the agent. A deterministic process is, for instance, the outage of a transmission line, when the power flow reaches a failure threshold. Stochastic processes are probabilistic component failure modes such as the unplanned outage of a power generator.

During the simulation, the agents provide informational input to the physical network layer (e.g., generator out of service due to maintenance), whose conditions are updated accordingly and then sent back to the agents which react to the new conditions in accordance with their behavioral rules. Regarding the modeling of the different agents, the principle is sketched by means of the TSO agent, as its explicit inclusion into the model is one of the main assets of ABM. In the model, the TSO is responsible for a certain control area of the overall power system.

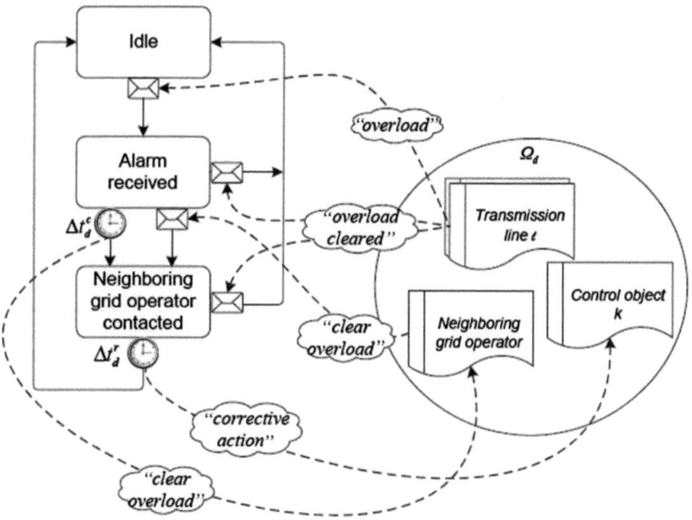

Fig. 6.23 Finite state machine: transmission system operator (TSO) agent

It becomes solely active in case of contingencies in which it tries to remove line overloads by re-dispatching the generators or shedding load if necessary. The FSM for the behavioral rules is illustrated for the overload of a transmission line connecting two control areas, see Fig. 6.23.

If the power flow measured by the transmission line agent becomes larger than a maximum allowable value, it sends an alarm message to the two responsible TSO agents. Upon receiving the alarm, the two neighboring TSOs have to contact each other: using ABM, such a 'human' behavior can be readily modeled by a time delay Δt_d^c. The operators then need some time to find a solution to the overload problem, which is modeled by a subsequent time delay Δt_d^r. During these time delays the overload might further increase and eventually trigger the outage of the transmission line, potentially resulting in a cascading blackout event. Otherwise, the corrective action to remove the overload is formulated as a conventional optimal power flow (OPF) problem. Thereby, the generator outputs are changed or load is shed in such a way that the flow on the line is reduced again below its maximum allowable value, while minimizing the costs of such a generator re-dispatch or of the potential load shedding.

The model has been applied to the three-area IEEE Reliability Test System 1996 in order to assess the sensitivity of the blackout frequency to an increase of the system loading. Therefore, the basic system demand and the generator capacities as given in (IEEE 1999) have been incremented by the same factor L, while keeping the transmission line capacities constant. Figure 6.24 shows the resulting complementary cumulative bclakout frequencies with respect to the unserved energy per event, $\hat{F}_c(C_E)$, for four different values of L (in normalized units). Regarding the two lower system loading levels ($L = 1.0$ and $L = 1.1$) the

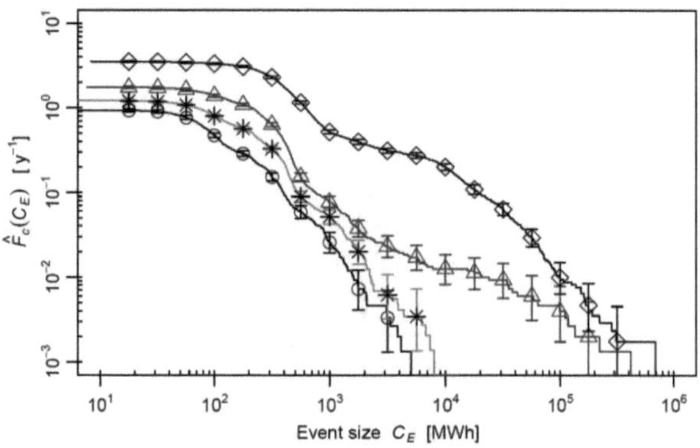

Fig. 6.24 Complementary cumulative blackout frequencies for four different system loading levels $L = 1.0, 1.1, 1.2$, and 1.37 (*circles*, *stars*, *triangles*, and *diamonds*) without operator intervention. The *error bars* indicate the 90% confidence interval

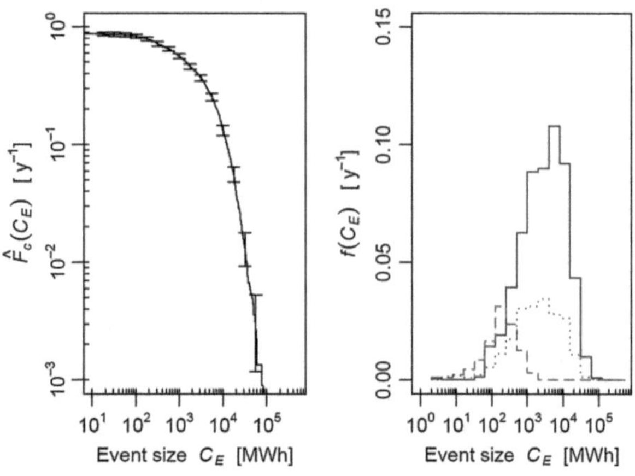

Fig. 6.25 *Left*: complementary cumulative blackout frequencies for the Swiss system with respect to the unserved energy. *Right*: histogram indicating the distribution of the outages due to generation inadequacy (*continuous line*), system splitting (*dotted line*), and load shedding for line overload removal (*dashed line*)

observed complementary cumulative frequencies follow approximately an exponential curve. However, increasing L to 1.2 already leads to a remarkable increase of large events, while the shape of the curve in the range of the smaller events (up to about 10^3 MWh) stays qualitatively the same. The value of $L = 1.37$ represents the maximum system loading level where no line overloads would occur without any stochastic component outages. This loading level can be characterized by a high frequency of large blackouts predominantly in the range between 10^4 MWh and 10^5 MWh. Hence, it can be concluded that the probability of cascading failures is subject to phase transitions and abrupt changes that result from only small changes in system stress.

The ABM technique has also been applied to the Swiss electric power system, in order to investigate its applicability for an in-depth modeling of a real system with respect to mid- and short-period power system planning purposes (Schläpfer et al. 2008). Selected results are depicted in Fig. 6.25, with respect to the unserved energy per event C_{E} (in particular, the complementary cumulative blackout frequencies, $\hat{F}_c(C_E)$, and the histogram of the different outage causes). The complementary cumulative blackout frequency follows an exponential curve.

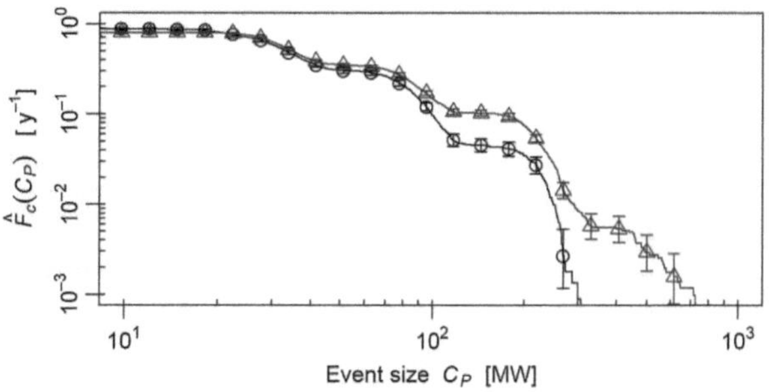

Fig. 6.26 Blackout prevention due to operator response to line overloads. *Triangles* no action, *circles* operator intervention with Δt_d^r=15 min

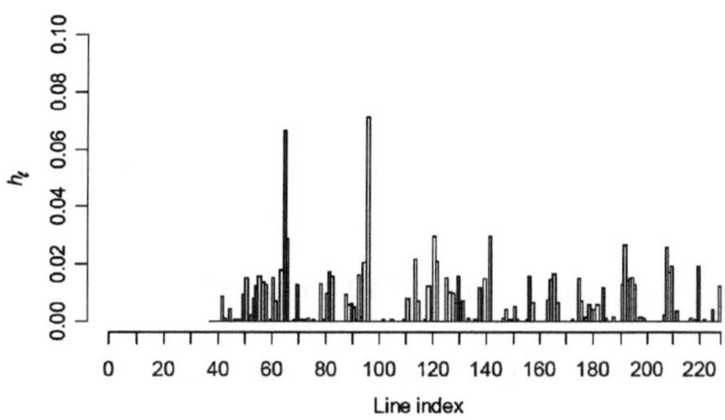

Fig. 6.27 Relative frequency of transmission line overloads

Generation inadequacy is the dominant factor regarding the larger events while load shedding for line overload relief becomes important in the range of the smaller events. The influence of load disconnections due to system splitting is significant but the frequency of this outage cause never exceeds the frequency of load disconnections due to generation inadequacy or load shedding due to the operator action.

Hence, under the corresponding model assumptions, it can be concluded that the reliability of the Swiss power grid is somewhat more sensitive to generation outages than to transmission line failures.

The impact of the operator response to a transmission line overload (Fig. 6.23) on the overall system reliability is shown in Fig. 6.26, comparing the frequencies of blackout events with and without operator response (i.e., $\Delta t_d^r = \infty$). In that case, the event size is measured by the maximum unserved demand per event.

The impact of the operator intervention becomes significant in the range of the larger events, where a high fraction of blackouts with a size C_P greater than 200 MW is prevented. These events need a high number of subsequently disconnected lines due to overload. Such a sequence, in turn, gives the operator a higher chance to intervene in comparison to the outage of few lines without further cascading.

As mentioned, agent-based modeling is suitable for the identification of critical components. As an example, the relative overload frequencies for each transmission line, h_l, are reported in Fig. 6.27. About 15% of all overload contingencies occur on only two lines. Furthermore, several groups of adjacent lines can be

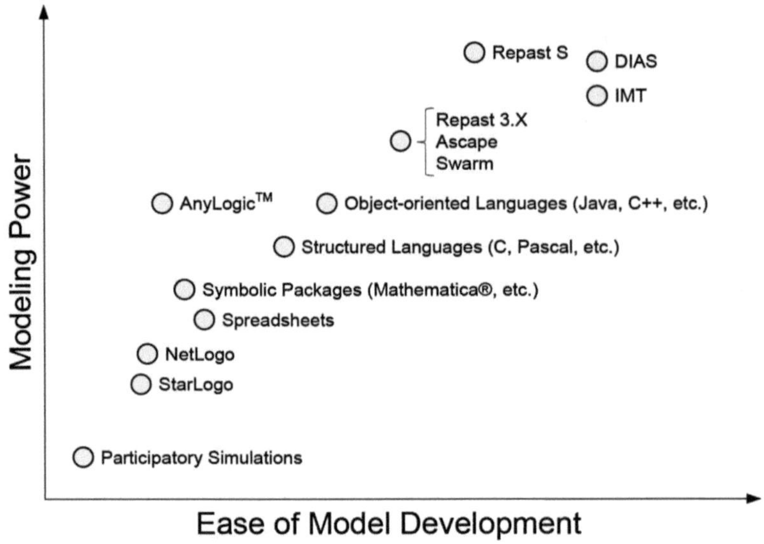

Fig. 6.28 Categorization of ABM tools (adopted from Macal and North 2005)

identified as being prone to overloads, helping to highlight the most critical regions of the system.

As for the computational efforts, in order to obtain statistically significant results for a system operating period of one year, around 50 h of simulation are needed on a single conventional desktop computer (Dell Optiplex GX260 with a Pentium 4 CPU of 2.66 GHz and 512 MB of RAM).

6.5.7 Available Software Tools

A plethora of software environments is available for supporting the technical implementation of the agent-based modeling concept. Figure 6.28 categorizes a number of widely used tools according to the ease of model development and the modeling power, respectively.

A listing of URLs for further information on selected tools indicated in Fig. 6.28 can be found below:

- StarLogo: www.media.mit.edu/starlogo
- NetLog: ccl.northwestern.edu/netlogo
- Mathematica: www.wolfram.com
- AnyLogic: www.xjtek.com
- Repast3.X: repast.sourceforge.net
- Ascape: ascape.sourceforge.net
- Swarm: www.swarm.org
- DIAS: www.dis.anl.gov/projects/dias

6.5.8 Conclusions

ABM offers an attractive modeling paradigm for describing the dynamic operational behavior of CI, with close adherence to the reality of the coupled processes involved. One of the major advantages is the possibility to include physical laws and time-dependent nonlinear phenomena into the simulation, and to emulate the behavior of the infrastructure as it emerges from the behaviors of the individual agents and their interactions. The level of modeling detail offered by ABM allows analyzing a multitude of time-dependent reliability and vulnerability aspects, e.g., system weak points and upgrades. Besides technical failures, other factors, such as natural hazards, institutional weaknesses or security-related issues can be integrated.

The main problems are related to the long computational time needed for the simulation and the large number of parameters to be input in the analysis. Moreover, properly quantifying the different types of uncertainties (see Chap. 5) further complicates the overall modeling procedure. However, by focusing on specific safety aspects, the model can be simplified and the computational burden reduced: gaining experience in applying the proposed approach is expected to give insights on the sensitive parameters to focus on.

6.6 High Level Architecture

6.6.1 Need for a Different Simulation Approach

When multiple interacting systems are planned to be represented in a single simulation tool, traditional simulation approaches often intend to integrate multiple simulation components in one simulation platform executing on one computer. This type of simulation approach apparently suffers from two key technical difficulties:

(1) *Lack of performance*: The increasing complexity of this type of simulation tools limits its performance, with consequences of continuous consumption of simulation hardware, increasing number of simulated systems, increasing demands for more accurate simulation validation, and increasing requests for more computational resources. This problem could be expected in any simulation tool developed through traditional approaches, and is further complicated when simulating interdependencies between CIs since more than one infrastructure need to be considered and more cross-infrastructure analyses needs to be conducted.
(2) *Lack of simulation interoperability*: According to the US Department of Defense (2007), simulation interoperability can be defined as “the ability of a system to provide data, information, services to and accept the same from

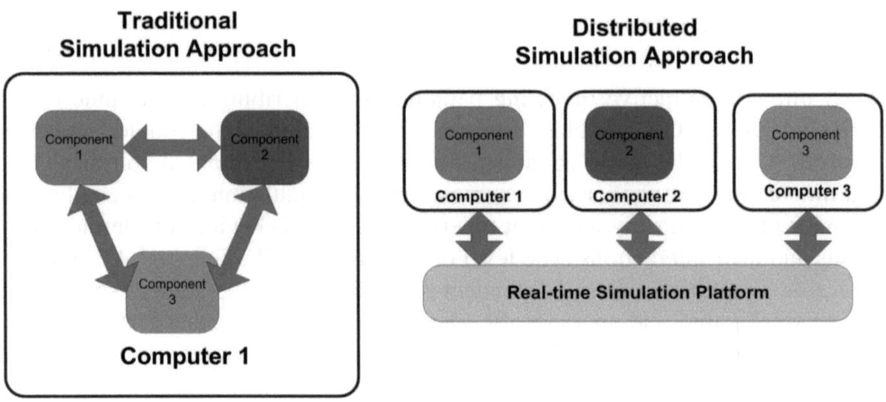

Fig. 6.29 Architecture comparison between two simulation approaches

other systems, and to use the data, information, and services so exchanged to enable them to operate effectively together". As the definition indicates, it is important to ensure effective data exchange capability between systems in order to improve simulation interoperability. However, the traditional simulation approach lacks this capability due to its inherent limitation, especially when it tries to simulate multiple systems in different domains, e.g., one system in time domain and another one in frequency domain.

One solution for these technical difficulties is to distribute different simulation components, which could be domain-specific or sector-specific across a simulation platform, so as to make the best use of computational resources. This approach, referred to as distributed simulation approach, can be considered as a successor of the traditional simulation approach in case multiple systems need to be simulated.[3] It changes the way to design and develop simulation tools, i.e., instead of building a "heavy weight" simulation component on one computer, a number of "light weight" components are developed on distributed computers interacting with each other over a real-time simulation platform, which not just potentially improves the efficiency and flexibility of the developed simulation tool but also decreases its overall complexity. Each distributed "light weight" simulation component is only developed to represent its own system characteristics. The architecture of this approach allows quick assembly of independently developed components without full knowledge of their peer simulation components.

The comparison between architectures of two approaches (traditional and distributed) is illustrated in Fig. 6.29. Benefits achieved from a distributed simulation approach can be demonstrated from an exemplary application for an aircraft simulation tool development. Suppose a newly designed navigation component is

[3] It should be noted that the distributed simulation approach should not be considered as an option if only one system (without any subsystems) needs to be simulated.

required to be tested with other components in this tool, before installing it on a real aircraft. It is not a good idea to develop a new aircraft simulation tool from scratch only for this purpose. Reusing existing component models with minor modification seems to be more promising and economic, which can hardly be accomplished using the traditional simulation approach. However, if this aircraft simulation tool is developed using the distributed simulation approach and all component models have been developed independently, tests can be easily performed since only navigation component model needs to be created or just modified.

While several simulation standards do exist for supporting the distributed simulation approach, the most widely implemented and applicable one is the high level architecture (HLA) simulation standard[4] (Pederson et al. 2006; Gorbil and Gelenbe 2009).

6.6.2 HLA Standard

HLA is a general purpose high-level simulation architecture/framework to facilitate the interoperability of multiple-types models and simulations (Pederson et al. 2006). Originally, HLA was developed under leadership of the US Defense Modeling and Simulation Office (DMSO)[5] for the purpose of supporting interoperation of simulations, reusing existing simulators for other purposes, and reducing the cost/time required to create a synthetic environment for a new purpose (Dahmann et al. 1997).

HLA baseline definition was completed in 1996. In April 1998, the first complete HLA interface specification was released to the public (DOD 1998). In 2000, HLA was approved as an open standard by the organization of the Institute of Electrical and Electronic Engineers: IEEE Standard 1516–2000 (IEEE 2000). Since then, the HLA standard has been revised and improved. The most current one is HLA-Evolved. One distinguished advantage compared to other simulation standards offered by HLA for the simulation industry is its support of live participants, meaning that the representation of the live world such as a human being, a real process instrumentation device or a controller, etc., can be integrated into the simulation world. Moreover, it is also capable to project data from simulation world back into real world (DOD 1998). A functional view of the HLA is given in Fig. 6.30.

As an open IEEE standard, HLA has been widely adopted across various fields of the simulation industry during the last decade. The EPOCHS (electric power

[4] Other simulation standards include distributed interactive simulation (DIS) and aggregate level simulation protocol (ALSP). However, their inherent weaknesses limit the capabilities as the standard for distributed simulation approach. For instance, ALSP is not able to support real-time communication and DIS fails to provide a time synchronization mechanism.

[5] DMSO has been renamed as US Department of Defense (DOD) Modeling and Simulation Coordination Office (M&S CO).

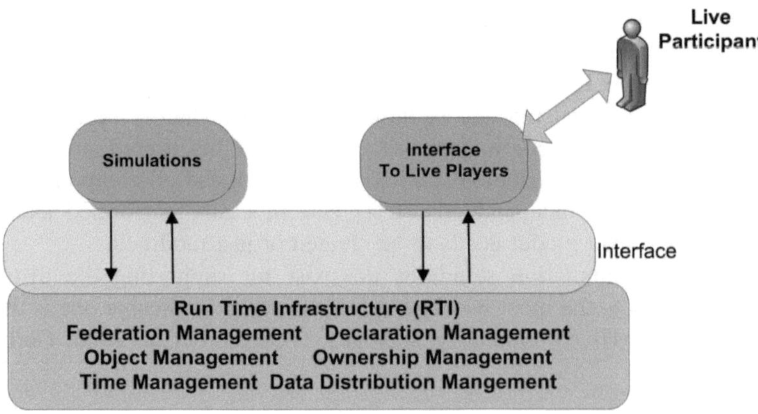

Fig. 6.30 Functional view of the HLA standard (Dahmann et al. 1997)

and communication synchronizing simulator) is an early attempt to distribute several individual simulators by adopting the standard of HLA, which utilizes multiple research and commercial systems from various domains (Hopkinson et al. 2003; Rehtanz 2003). An HLA-based system for geo-computation in a grid environment was designed and developed at Inha University of Korea for the purpose of CDM (communication data and management) performance evaluation (Kim et al. 2006a). Computer experiments conducted by Lees and Logan show that "the overall simulations have been sped up after distributing simulation components based on the standard of HLA" (Lees et al. 2007). Similar results are also observed by Zhao while working on an agent framework for controlling the activity flows between the ISS (interactive simulation systems) components (Zhao et al. 2005). Furthermore, HLA has been applied to other industry fields, such as US border operation study (Beeker and Page 2006), rail traffic safety system simulation (Lieshout et al. 2008), and many others (Ezel 2007; Möller et al. 2005; Zacharewicz et al. 2009).

Adopting HLA for the CI (inter)dependency study is also not a new concept. In 2007, HLA approach has been considered an interface solution for trying to connect several individual simulators to study interdependencies between heterogeneous interconnected CIs (Duflos et al. 2007). In 2009, a communication middleware serving other distributed CI simulators was created by a team in a EU research project "Design of an Interoperable European Federated Simulation Network for Critical Infrastructures (DIESIS)" (Gorbil and Gelenbe 2009). This middleware, adapted from the HLA standard, aims to provide a reliable one-to-one real-time communication platform for diverse simulators over the WAN (Wide Area Network). Currently, an HLA-compliant experimental simulation test-bed for the purpose of studying (inter)dependency between SCADA (supervisory control and data acquisition) system and EPSS (electricity power supply system) is under development at ETH Zurich (Eusgeld and Nan 2009).

Table 6.12 Federate/federation rules (IEEE 2000)

Federation rules		Federate rules	
Rule	Rule description	Rule	Rule description
1	Federations shall have an HLA FOM (federation object model), documented in accordance with the HLA OMT (object model template)	6	Federates shall have an HLA SOM (simulate object model), documented in accordance with the HLA OMT
2	In a federation, all simulation-associated object instance representation shall be in the federates, not in the RTI (run time infrastructure)	7	Federates shall be able to update and/or reflect any instance attributes and send and/or receive interactions, as specified in their SOMs
3	During a federation execution, all exchange of FOM data among joined federates shall occur via the RTI	8	Federates shall be able to transfer and/or accept ownership of instance attributes dynamically during a federation execution, as specified in their SOMs
4	During a federation execution, joined federates shall interact with the RTI in accordance with the HLA interface specification	9	Federates shall be able to vary the conditions (e.g., thresholds) under which they provide updates of instance attributes, as specified in their SOMs
5	During a federation execution, an instance attribute shall be owned by at most one joined federate at any given time	10	Federates shall be able to manage local time in a way that will allow them to coordinate data exchange with other members of a federation

Generally, HLA consists of three essential elements:

(1) *Federate/federation rules*: Defined by the HLA standard, each distributed component is referred to as a federate and the collection of federates that comprise a simulation is referred as the federation. A set of 10 HLA rules that the federation and all participant federates must follow are defined by the standard IEEE1516-2000 to be considered as HLA compliant. These rules can be grouped into a set of five rules for HLA federates and five rules for the federation, both shown in Table 6.12.

(2) *Object model template (OMT)*: All objects and interactions implemented by a federate should be visible to all other participant federates across the federation, if necessary, to guarantee the interoperability between federates. Therefore, they must be specified in detail with a common format. OMT provides a standard for declaring corresponding information of two HLA object models: the HLA federate object model (FOM) and the HLA simulate object model (SOM), which have been mentioned in Table 6.12. FOM describes the set of objects, attributes, and interactions shared by all federates under one federation. SOM describes all objects, attributes, and interactions that one federate can offer. One federation only requires one FOM and each federate must have one SOM.

(3) *Interface specification*: The HLA interface specification identifies how federates interact with the federation, as well as with each other and is

Table 6.13 HLA runtime services (IEEE 2000)

Runtime service	Purpose of the service
Federation management	Create, operate, and remove a federation
Declaration management	Declare what information a federation will offer and require
Object management	Provide services, such as creation, deletion and identification at the object level
Ownership management	Manage ownership of objects/attributes of all federates
Data distribution management	Route data transmission among federates during federation execution
Time management	Synchronize time among federates during federation execution

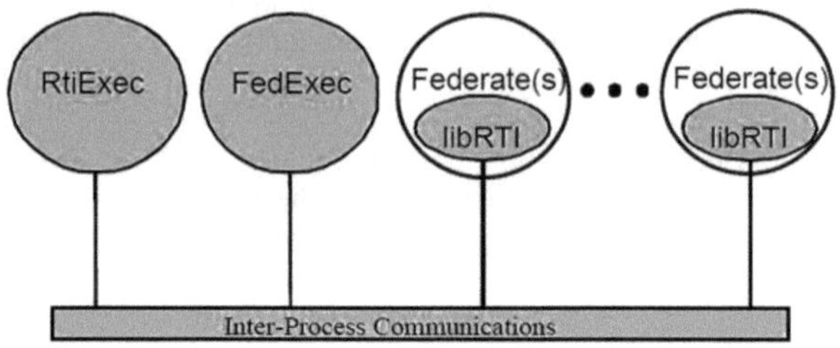

Fig. 6.31 Three major RTI components (DOD 2000)

implemented by the RTI (run time infrastructure) during the federation execution. The HLA interface specification defines runtime services provided to federates by the RTI, and by federates to RTI. Six runtime services are specified by the HLA interface specification and a list of these runtime services is shown in Table 6.13.

6.6.3 Run Time Infrastructure

While HLA is an architecture, a simulation standard but not a software, RTI is the software. It is the core element of the HLA standard providing common services to all federates. Interactions between federates in a federation, as well as between federates and federation, are all accomplished via the RTI. Generally, RTI consists of three major components, showed in Fig. 6.31 (DOD 2000).

- *RtiExec*: A global known process that manages the creation and destruction of federation execution.
- *FedExec*: A federate-based process that manages federates joining into and resigning from the federation.
- *LibRTI*: A C++ or Java library that provides all RTI services for developers, defined by the HLA interface specification.

Major interplays between a federate and its joined federation, defined by the HLA interface specification and implemented by the RTI, are shown in Fig. 6.32,

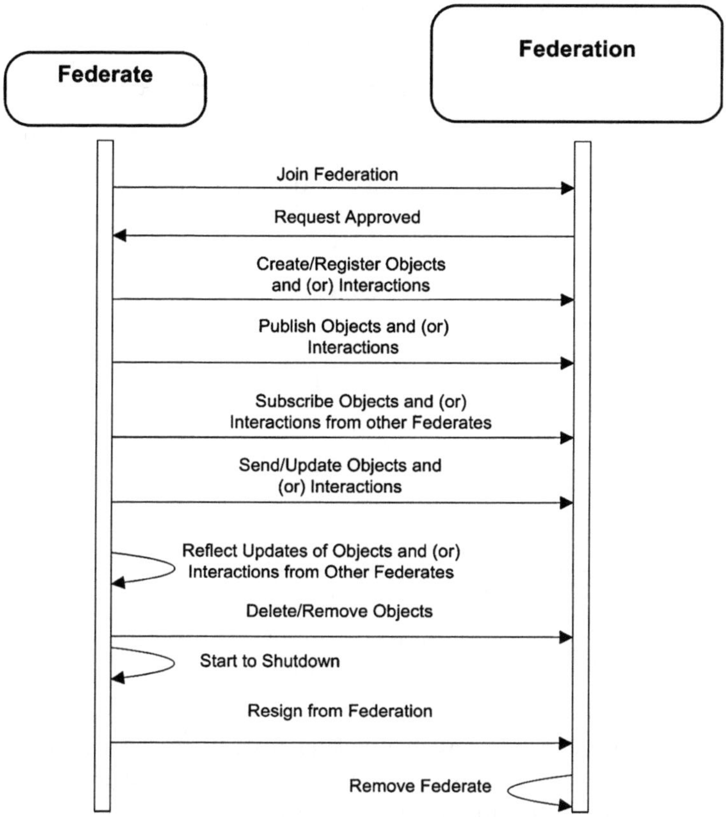

Fig. 6.32 Major Federate–Federation interplays (DOD 2000)

which is a modified figure based on the work done in (DOD 2000). If a federate attempts to join an existing federation and become a participant federate, a "join" request must be sent to the federation. After receiving the approval response from the federation, it becomes a participant federate and must publish/subscribe corresponding object and interaction classes. Specified by the HLA standard, both object class and interaction class can be used to define a possible empty set of named data, which are called attribute and parameter, respectively. The purpose of these two classes, which are also called HLA-related classes, is to store the data which will be transmitted between federates. The only difference between two classes is that the interaction class will be destroyed after its contained data have been received.[6] While the purpose of publishing HLA-related classes by a federate

[6] HLA standard does not describe which class should be implemented during the simulation development, which is a developer's task to decide according to simulation requirements. It is possible that both classes are implemented or only one class is implemented, in one federate.

Table 6.14 Comparison of several RTI software tools

	Pitch pRTI™	MÄK RTI™	Portico RTI	CERTI
Type	Commercial	Commercial	Open (free)	Open (free)
Supported HLA standard(s)	HLA 1.3, IEEE (HLA) 1516, HLA Evolved	HLA 1.3, IEEE (HLA)1516, HLA Evolved	HLA 1.3	HLA 1.3
LibRTI Language	C++, Java	C++	C++, Java	C++
Software support?	Yes	Yes	No	No
Console interface included?	Yes	Yes	No	No
Maxim number of federates supported	>10	>10	Limited	Limited
Continuous developments?	Yes	Yes	NO	Yes
Web communication supported?	Yes	Yes	NO	NO

Pitch pRTI™: www.pitch.se (2010)
MÄK RTI™: www.mak.com (2009)
Portico RTI: http://porticoproject.org (2010)
CERTI: http://savannah.nongnu.org/projects/certi/ (2010)

is to inform other federate(s) possible updates from these classes, the purpose of subscribing HLA-related classes by a federate is to inform other federate(s) what classes it (federate) would like to receive updates from. During the simulation, whenever published HLA-related classes of the federate are updated, the updated data will be broadcasted to all available federates across the federation. However, only federates who have previously subscribed these classes will be able to receive the updated data. If the federate attempts to quit the joined federation, all its owned HLA-related classes must be deleted/removed from the federation before sending a "resign" request to the federation. As soon as this request is received by the federation, the federate will then be removed from the federation. Although there are many other important federate-federation interplays such as federate time synchronization, time management, HLA-related classes ownership management, etc., they all belong to advanced topics of the HLA standard and are not subjects of this book.

Developing RTI software is a complicated and tedious task. Although all classes and methods have been well defined and described by the HLA standard, the implementation is not easy. For example, it took several software engineers about 3 years to complete the first public version of a RTI software tool, which is called Portico RTI. Therefore, developing own RTI software is not recommended. A list of ready-to-use RTI software tools is shown and compared in Table 6.14. A list of URLs for further information on these tools can also be found below:

6.6.4 Recommended Work Steps

Adopting HLA as a standard to develop the distributed simulation approach can be divided into the following steps:

- *Step 1—Feasibility study*: Not all simulation components are able to be distributed using the HLA standard. A "pre-screening" investigation is highly recommended before considering the HLA standard as an option. Whether or not distributing simulation components, which means breaking down their interlinked functions, will affect the final outcomes of the overall simulation is the main concern. The feasibility of distributing simulation components, especially when modeling multiple subsystems existing under one system, should be carefully studied and verified. More details related to this step can be found in Eusgeld and Nan (2009).
- *Step 2—RTI software tool selection*: The following questions can be used to steer the decision on which RTI software tool should be selected:
  - Which HLA standard is required or preferred by the developers?
  - What is the major programming language for developing distributed components (e.g. Java or C++)?
  - How many federates are planned to be developed?
  - Is Web-supported RTI software tool required for the development?
  - Is it necessary to reuse any existing simulators?
  - Is RTI software support from vendor necessary?
- *Step 3—object/interaction class definition*: For a federate, it is important to determine which variables will be updated and which will be of interest for other peer federates. Then, HLA related classes (object/interaction class) can be precisely defined and implemented, which is an essential step for FOM definition.
- *Step 4—local RTI interface development*: After the RTI software has been selected, the local RTI interface that contains the classes and methods used to connect to the federation must be implemented for each federate. All HLA-related functions that are responsible to exchange data between federates are conducted by the local RTI interface. As part of the federate, the local RTI interface should be developed in the same programming language used to develop the federate. Implementing the local RTI interface for a previously non-HLA-compliant simulation component needs to be conducted carefully, since any mistake during the modification could result in the failure of the whole component.
- *Step 5—federation tuning*: Federation tuning is the last step to implement the HLA-compliant simulation platform, which will help to improve the efficiency and accuracy of the overall simulation. Several capabilities, such as simulation interoperability, time synchronization, and data exchange rate, provided by the HLA standard can be studied and then improved by modifying the corresponding simulation parameters to optimize the simulation performance.

6.6.5 Drawbacks of the HLA Standard

The major drawbacks of the HLA simulation standard are:

- *Significant increases of resources and time during implementation*: Resources and time required to implement an HLA-compliant simulation platform could be significant comparing to non-HLA-compliant simulation platforms. This is mainly caused by the development of a local RTI interface component for each federate.
- *Update latency*: Update latency, which means the interval between sending an update by one federate and receiving this update by another federate, could be significant enough to affect the outcomes of real-time simulation. It should be noted that negative effects of this drawback can be alleviated by improving/upgrading the hardware environment of the simulation platform such as by using computers equipped with a better CPU.
- *Not a "plug-and-play" standard*: All HLA-related (object and interaction) classes must be declared in advance before the simulation. As a consequence, adding or even modifying these classes becomes impossible during the simulation.
- *Incompatibility between HLA standards*: An example of this drawback is that a federate developed based on standard of HLA 1.3 is not able to join the federation developed based on standard of HLA1516, unless being upgraded to HLA1516; this means that any future changes to the HLA standard may have significant impact on local implementations.

6.6.6 Exemplary Application

The HLA-compliant experimental test-bed, which is part of an ongoing broader-scale project in the area of CI vulnerability and (inter)dependency studies at ETH Zurich, is an exemplary application of HLA and has been mentioned in Sect. 6.6.2. The test-bed recreates the architecture of a typical EPSS with its own SCADA system to investigate and study hidden vulnerabilities between two systems due to their (inter)dependencies. Originally, the experimental test-bed was intended to be developed through a traditional simulation approach but several technical issues arose during the feasibility study, such as the difficulty of reusing a simulation component which has been developed for another project by means of a different simulation tool, computational capability limitation of the available computers, possibility of using the test-bed for other research projects, etc. Thus, the developers decided to build the test-bed by distributing simulation components according to the HLA standard on different computers over the local area network (LAN). Currently, the test-bed consists of three major simulation components: EPSS simulator, SCADA simulator, and central RTI. The architecture of the test-bed is illustrated in Fig. 6.33 and summarized below.

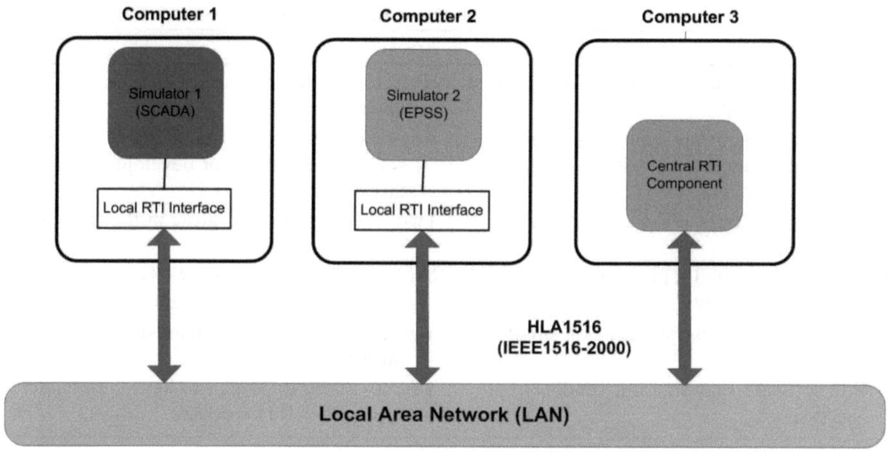

Fig. 6.33 The architecture of the HLA-compliant experimental test-bed

- *Central RTI component*: It acts as the centre of the experimental test-bed. This component is responsible for simulation synchronization and communication routing between all components, through local RTI interface of each simulator. The central RTI is a globally known component. Each federate communicates with the central RTI via its own local RTI interface and starts to follow central federation management.
- *EPSS simulator*: This component is a time-stepped and object-oriented simulator, which has been developed using the software of Anylogic 5.5. Originally, as discussed in the previous section, the EPSS simulator was a stand-alone simulator developed to integrate stochastic time-dependent technical and non-technical factors into a vulnerability assessment based on a two-layer object-oriented modeling approach (Schläpfer et al. 2008). This simulator has been introduced in Sect. 6.5 as an exemplary application of object-oriented modeling.
- *SCADA simulator*: This component is an event-driven and object-oriented simulator, which has been developed using the software of Anylogic 6.4. The development of this simulator aims at studying and investigating hidden vulnerabilities between SCADA and its monitored/controlled EPSS due to (inter)dependencies. The SCADA simulator has been developed to be able to:

  (1) Acquire information sent by the EPSS simulator.
  (2) Analyze the collected information according to predefined scenarios.
  (3) Send commands to the EPSS simulator based on the results from data analysis.

The HLA-compliant experimental test-bed has been developed following the five steps introduced before:

- *Step 1—feasibility study*: In this project, the main purpose of the SCADA system for the EPSS is to allow an operator or user to collect data from distant

Table 6.15 Answers for RTI software tool selection investigation

Question	Answer
Which HLA standard is required or preferred for developers?	HLA 1516 is the preferred HLA standard by all developers
What are the major programming language developing models of distributed components?	Anylogic, a Java-based model development software, is the major development tool in this project
How many federates are planned to be developed?	Two (this number will grow in the future)
Is the Web-supported RTI software required for the development?	No, but a Web-supported RTI software is preferred by all developers
Is it necessary to reuse any existing simulators?	Yes, one non-HLA-compliant simulator (EPSS) must be used
Is RTI software support from vendor necessary?	Yes, it is very important to have continuous support for RTI software

electricity transmission substations and send control commands in case of detection of deviations from normal working conditions. Interlinked functions, meaning the functions that involve both systems, are present due to the functional interconnections between the two systems. Distributing these two simulation components, representing the corresponding systems, indicates that all interlinked functions must be broken down. It is very important to ensure that the decomposition of the interlinked functions will not affect the accuracy of the overall simulation results, which has been carefully studied and proved by the developers of this project. More details regarding this feasibility study can be found in Eusgeld and Nan (2009).

- *Step 2—RTI software tool selection*: In order to choose an appropriate RTI software tool, the list of questions introduced in the previous Section has been studied with answers shown in Table 6.15. Based on those answers, software tool pRTI™ from Pitch Technology has been selected. pRTI™ is the leading HLA run time infrastructure for the international IEEE 1516 standard, certified by DMSO in 2003, and is now used by thousands of customers in major high-tech companies all over the world. More information regarding pRTI™ and Pitch Technology can be found from www.pitch.se.
- *Step 3—object/interaction class definition*: Distributed simulation components should be capable of representing interconnections between the two studied systems (SCADA and EPSS). An example of this type of interconnection is that the SCADA system requires the measured process variable, which is the output of EPSS, and on other hand, EPSS requires the most recent operating status of the field control device, which is the output of SCADA system. Descriptions of several object classes already defined in this simulation development are shown in Table 6.16.
- *Step 4—local RTI interface development*: Local RTI interface can be implemented by inheriting and modifying corresponding classes and methods from the RTI software tool pRTI™. As discussed before, EPSS simulator was previously designed as a stand-alone simulator, no inputs from external simulators

Table 6.16 Descriptions of several object definitions

Object	Attribute	Type	Federate (Simulator)	
			SCADA	EPSS
Transmission line	measured Variable	Double	subscribe	publish
	status Variable	Boolean	publish	subscribe
	control Command	Integer	publish	subscribe
Generator	power Generate	Double	subscribe	publish
	power Inject	Double	subscribe	publish
Load	actual Load	Double	subscribe	publish
	demand Load	Double	subscribe	publish

have been specified. To include this simulator in the test-bed, it has been revised as an HLA-compliant simulator by adding an independent local RTI interface without any modification of the simulator.

- *Step 5—federation tuning*: A number of experiments analyzing simulation performances have been designed and conducted after setting up the experimental test-bed. For example, one experiment is especially developed to study the data exchange rate between federates and how the hardware configuration of each simulation component (federate) will affect the overall simulation. Based on the results from this experiment, the component hardware and several HLA-related parameters have been modified to maximize data exchange rate between federates and optimize the simulation performance.

Currently, the architecture of the test-bed has been successfully created. Although the SCADA simulator is not yet fully implemented, experiments have been designed and conducted on the test-bed with the results demonstrating the capability and applicability of the HLA as a simulation standard for implementing a distributed simulation approach. Figure 6.34 illustrates the simulation results from an experiment designed to study the negative effects of an accidentally overloaded transmission line, which is also summarized in Table 6.17. In this experiment, it is assumed that whenever a monitored transmission line is overloaded, an alarm will be generated and sent to the power system operator in the control centre by the remote terminal unit (RTU)[7] of the SCADA for the purpose of notification. If after a timeout, the operator fails to react to the overloading alarm, then the protection devices such as the line disconnector will automatically disconnect the overloaded transmission line to minimize negative consequences. To observe three different outcomes after the occurrence of the transmission line overload, three case study scenarios have been developed by modifying the parameters of the corresponding agents (the operator, protection device and transmission line) in two simulators (EPSS and SCADA). More details regarding the illustration and analysis of this experiment can be found in Nan and Eusgeld (2011).

[7] RTU is the remote device of SCADA, which is usually located away from the control centre and is responsible for acquiring physical data from the field and executing the control instruction(s) sent from the control centre.

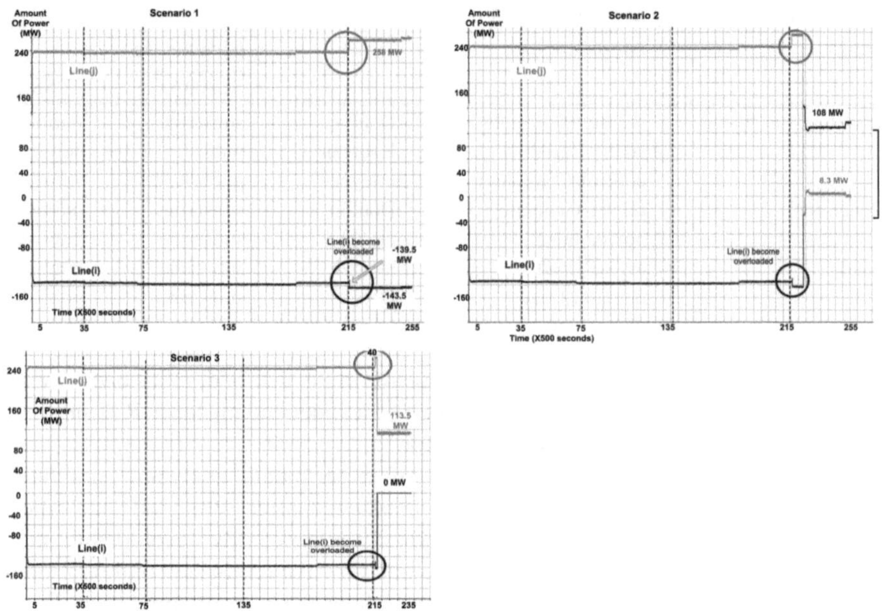

Fig. 6.34 Simulation results of the transmission line overloading experiment in (Nan and Eusgeld 2011)

Table 6.17 Summarized simulation results of the transmission line overloading experiment in (Nan and Eusgeld 2011)

Scenario	Failure of the operator to react	Failure of the protection device	Observed results
1	Yes	Yes	Power of interconnected line (line(*j*)) starts to increase
2	No	N/A	Power of overloaded line (line(*i*)) starts to drop
3	Yes	No	Power of overloaded line (line(*i*)) drops to zero

N/A = not applicable

6.6.7 Conclusions

Studying and analyzing the vulnerabilities of CIs particularly related to (inter)dependencies, often involve the development of multiple simulation components representing the characteristics of each studied infrastructure, and their integration into a single simulation tool. Compared to the traditional simulation approach, which attempts to build multiple simulation components into one simulation tool executing on one computer, a different (distributed) simulation approach can be undertaken, which distributes all developed components geographically and links them together into one simulation platform.

HLA is an open simulation standard supporting such a simulation approach and aims at improving the interoperability and efficiency of simulations composed of different simulation components. As an open simulation standard, HLA has been involved in a large number of simulation development works for various purposes. Most importantly, it has been integrated into projects sponsored by different governments. For example, a simulation platform based on the standard of HLA1516 (IEEE1516-2000), which includes the components of pilot, air traffic controller, weapons controllers, and fighter allocators, has been developed by the Swedish Air Force Air Combat Simulation Centre (FLSC), part of the Swedish Defense Research Agency, to provide training services mainly for the Swedish Air Force (Möller et al. 2005). A similar simulation platform has also been developed by the National Aerospace Laboratory of Netherlands (Lemmers et al. 2002).

Although the HLA standard has several disadvantages, it has been continuously improved during the last decade benefiting from its widespread international acceptance. As a consequence, the robustness and performance of recently implemented RTI software tools have been enhanced considerably. Together with improvements of CPU technology, the RTI software tools implemented according to the newest HLA standard, HLA Evolved, are able to provide 50,000 (sometimes 100,000) updates of 100 bytes per second between two federates over a LAN with update latency of less than 130 ms (Morse et al. 2006; Möller et al. 2008; IEEE 2009). Other benefits include modular FOM support, improved fault tolerance, individual federate update rate configuration, RTI debugging, etc.

Developing such an HLA-compliant distributed simulation platform is not an easy task. Furthermore, not all simulation tools are able to be developed using the HLA standard. The feasibility and applicability of distributing simulation components should be carefully investigated and verified before the implementation. To utilize the maximum capacity offered by the HLA standard, several work steps have been recommended. The major concern of adopting the HLA standard is the significant increase of resources and time needed for HLA-related application and interface developments. However, after integrating the HLA standard into the simulation platform, more benefits can be expected for future developments, e.g., improved flexibility and modularization of simulation development, distribution of simulation work load, and possibility of reusing models/simulators from other simulation platform.

6.7 Human Reliability Analysis

Human reliability analysis (HRA) attempts to estimate the likelihood of particular human actions (that may prevent hazardous events) not being undertaken when needed, or other human actions that may cause hazardous events (by themselves or in combination with other conditions) occurring (Wreathall et al. 2003). The main objectives of HRA are:

(1) To ensure that the key human interactions are systematically identified, analyzed and incorporated into the safety analysis in a traceable manner.
(2) To quantify the probabilities of their success and failure.
(3) To provide insights that may improve human performance.

Any attempt at estimating the likelihood needs to consider the work environment and the task conditions under which the work is done, as these can be an important influence on the likelihood of error. For example, bad environmental conditions, fatigue, stress or high workload. In turn, the work environment and task conditions are often influenced by organizational factors such as work rules, safety culture or training. Therefore, the error estimation process needs to account for all of these contributing factors.

HRA uses qualitative and quantitative techniques to assess the human contribution to risk, in particular the likelihood of required human actions being performed when needed (Bell and Holroyd 2009) These likelihoods can then be incorporated into the overall risk assessment, and combined with other probabilities, such as those of equipment faults, to estimate the overall likelihood of hazardous events. By adopting a structured and systematic approach to the assessment of human performance, HRA can provide a high degree of confidence that the safety and availability of complex technological systems, including Critical Infrastructures (CIs) are not unduly jeopardized by human performance problems (Kyriakidis 2009a, b).

6.7.1 Critical Infrastructures and HRA

Chapter 2 introduced the notion of CIs to describe networked systems and assets that are essential for modern societies and economies. Threats to a CI can be manifold; this sub-chapter is concerned with the analysis of techniques to assess human, caused threats.

Since the middle of the last century, perspectives on the design, operation, and maintenance of technological systems have significantly altered. Technological development had reached a state whereby the capabilities of the unaided human increasingly became the limiting factor in the performance of the overall system. To overcome this, human factors were taken into account in the design of systems to ensure that the demands on human performance did not exceed the natural capabilities of humans. Since the 1970s, HRA has become highly developed to improve the study of human influence and contribution to a system's reliability (Kyriakidis 2009a, b), developing techniques that evolved from first- to second-generation HRA.

First-generation tools were developed to help risk assessors predict and quantify the likelihood of human error. These include pre-processed tools and also expert judgment approaches. Characteristics of first generation approaches are that they tend to be atomistic in nature and encourage the assessor to break a task into component parts prior to considering the potential impact of modifying factors, such as time pressure, equipment design, and stress.

By combining these elements the assessor can determine a nominal human error probability (HEP). First generation techniques focus on the skill and rule based level of human action and are often criticized for failing to consider such factors as the impact of context, organizational factors and errors of commission (Bell and Holroyd 2009). Despite these criticisms, they are useful and many such techniques are in regular use for quantitative risk assessments (QRAs).

The 1990s saw the development of second generation techniques, a process that is on-going. Such tools attempt to consider context and errors of commission in human error prediction. However their widespread use has been slow to say the least, and consequently the benefits of the second generation over first generation approaches are yet to be established. Furthermore, such techniques have yet to be empirically validated. Literature highlights that second generation techniques are generally considered to be still under development but that in their current form they can provide useful insights into human reliability issues (Bell and Holroyd 2009).

This chapter will critically assess some of the most well known HRA techniques in the following four domains of CI:

(1) Transport, in particular railways, aviation, and the transport of dangerous goods by road
(2) Energy, including electrical network
(3) Public health
(4) Information and communication technologies

In each of these domains, an example will be given with a known HRA technique to illustrate the process of utilizing the technique. In addition, this chapter will identify areas for further investigation. The authors will not provide a comprehensive review of HRA techniques as many such reviews exist (Bell and Holroyd 2009) nor do they claim that these techniques are the only ones to use in CIs; rather they attempt to provide the reader with an introductory knowledge of the HRA techniques and their applications. In addition, the authors note that the selection of the HRA techniques presenting in this chapter was based on whether there is an application experience of the method in the corresponding CI domain.

Not all the methods are described in this chapter have the same degree of maturity. Two of them, the THERP and the HEART have been empirically validated and applied to several domains. On the other hand CARA has been applied only to air traffic control domain (Bell and Holroyd 2009), while CREAM is applied for the first time to a scenario related to an electrical network (Kyriakidis 2009b). Finally SHERPA, has been applied to healthcare industry, but it could also be applied to other CI domains. Table 6.18 illustrates a summary assessment of the techniques which will be described in detail in the next sections.

Table 6.18 Summary of HRA techniques

HRA technique	Advantages	Disadvantages	Comments
THERP	1. It is an overall, well used in practice methodology 2. It offers a powerful methodology, which can be made auditable by the assessor 3. It is performed well in terms of accuracy	1. It is relatively unstructured 2. It can be time consuming 3. It is highly judgmental based on assessor's experience 4. Its interaction between certain PSFs is unknown, therefore, can be given no guidelines for possible combinations	1. First-generation technique 2. Quantitative and qualitative results 3. Designed for nuclear industry 4. Generic tool applied to offshore, medical, transport domains 5. Training required on the method
SLIM	1. It is a flexible technique and a good theoretical method 2. It is able to deal with the total range of human error forms 3. It does not need task decomposition (task analysis and error taxonomies)	1. It is a complex method that needs intensive resource 2. The choosing of PSFs is quite arbitrary 3. It is a subjective method, something that reduces its reliability and consistency 4. Some times there are problems regarding experts' group synthesis 5. There is lack of valid calibration data	1. First-generation technique 2. Designed for nuclear industry 3. Applied to nuclear power plants and chemical industry 4. Suitable for application in major hazard domains 5. Requires an expert panel to conduct the assessment
ATHEANA	1. It is a focused prediction of the specific error that might be made and the most influential factors affecting that specific error 2. It increases assurance that the major risk associated with the HFE has indeed been captured 3. It is able to estimate HEPs for all sorts of combinations of factors and various conditions 4. It considers more PSFs than other techniques 5. It increases the guarantee that the key risks associated with the HFE in question have been identified	1. The primary shortcoming of the technique is that from a PRA stance, there is no HEP produced. As a result, the ease with which this analysis can be fit into a predictive quantitative risk assessment is reduced 2. It fails to prioritize or establish details of the causal relationships between these factors. Thus, further work is required to be performed in order to establish the root causes of an incident from a HRA perspective 3. The outcomes of the human errors under consideration are constrained by previously defined sequences of PRA accidents	1. Second-generation technique 2. Qualitative and quantitative results 3. Designed for nuclear industry 4. Can be applied to other domains 5. Training required on the method
CARA	1. It provides an initial indication that human reliability analysis can be used to deal with human factors arguments in a quantified ATM safety case context 2. It is a quick and simplistic approach 3. It gives a quantitative and qualitative output 4. It uses 6 generic type tasks descriptions	1. It has been applied to limited cases 2. It can be time consuming when new tasks are introduced 3. Neither dependence nor EPCs interaction is taken into account	1. First-generation approach 2. Quantify human reliability aspects as failure rates and success of mitigation actions in ATM 3. Designed for aviation domain
HEART	1. It is a quick and simplistic approach that needs little of training 2. It gives error reduction suggestions 3. It gives the analyst a quantitative output 4. A number of validation studies have produced encouraging results	1. It is a subjective method, something that reduces its reliability and consistency 2. Neither dependence nor EPCs interaction is taken into account 3. Little guidance is offered to the analysts in a number of the key HEART stages, such as the assignment of EPCs, or their exact number is included in a scenario	1. First-generation technique 2. Quantifying human error. 3. Applied to nuclear, chemical, aviation, rail, and medical industries 4. Can be applied to all major hazards sectors

(continued)

Table 6.18 (continued)

HRA technique	Advantages	Disadvantages	Comments
			5. Generally easy to be applied by engineers and human factors specialist
CREAM	1. It has the potential to be extremely exhaustive 2. The context is considered when using CREAM 3. It is a clear, well structured, schematic and systematic approach to error identification and quantification 4. It can be used proactively and retrospectively as well as in several domains 5. Its classification scheme is detailed, exhaustive and takes into consideration also environment and system causes of error	1. It appears complicated and daunting for a novice 2. The exhaustiveness of the classification scheme (error taxonomy) serves to make the method more time and resource intensive than other techniques 3. It has not been used extensively 4. It does not offer any remedial measure (ways to recover human erroneous actions are not provided) 5. Even for very basic scenarios the necessary time for applying CREAM would be high 6. It would presumably require analysts with knowledge of human factors and cognitive ergonomics, thus analysts with less experience might face difficulties on applying it	1. Second-generation technique 2. Can be applied for both retrospective analysis and performance prediction 3. Applied to nuclear, rail and aviation industries 4. Can be applied to other major hazards sectors 5. No special skills or knowledge are required but human factors knowledge would be advantageous for its use
SHERPA	1. It is a structured and comprehensive procedure, yet maintains usability 2. Its taxonomy prompts analysts for potential errors 3. It encourages validity and reliability data 4. In addition to predicted errors it offers error reduction strategies as part of the analysis 5. It can be used to analyze tasks or processes at many different levels 6. It can be adapted to different ward settings 7. It can be applied to a different range of health-care procedures	1. It can be tedious and time-consuming for complex tasks 2. An extra work for the analyst is necessary if HTA is not already available 3. It does not include performance-shaping factors 4. Does not model cognitive components 5. It is not a quantification technique 6. The task analysis has to be drawn up before the error predictions can be made 7. To gain a full description of every step of the healthcare task, several long HTAs are required 8. In cases-tasks with no formal protocol could be time-consuming to achieve a high level detail 9. It does not include in the analysis cases that are supposed unlikely to be happen 10. It is not the case that SHERPA taxonomy is able to capture the full range of error producing activity. Communication with patients and their relatives, colleagues, and various departments all impact on the process of the task, e.g. drug administration. These factors cannot be analyzed effectively using the existing taxonomy and would require other techniques (Lane et al. 2006) or a updated version of SHERPA	1. Originally designed for process industries 2. Applied to medical and aviation industries 3. Can be applied to other major hazards domains 4. Training on the method is recommended

6.7.2 *Transport Domain*

Whether it is railways, aviation or road transport, humans play a major role in their safe operation. The techniques highlighted below are chosen as they are both a good representation of common approaches to HRA as well as being sufficiently documented and discussed in the literature. However, it should be noted that they are not the only techniques that can be applied to CIs. These techniques were initially developed for application in the nuclear domain, though subsequently they have had a wider applicability (Eurocontrol 2007; Bell and Holroyd 2009).

6.7.2.1 Railways

Evidence indicates that human error was involved in 70% of railways accidents in the UK between 1900 and 1997 (Wreathall et al. 2003). There have been tentative developments in HRA techniques in recent years, especially in the UK and USA. The main areas of interest have focussed on driver and signaller's tasks. The example below involves the use of a HRA technique for train drivers.

HEART

The human error assessment and reduction technique (HEART) is a first generation HRA technique, which offers an approach for deriving the numerical probabilities associated with error occurrence. HEART was designed as a quick, easy to use and to understand HRA method, and is a highly structured approach that allows analysts to quantify potential human error. One of the major features of this approach is that in order to reduce resource usage, it only deals with those errors that will have a main impact on the system in question. The method uses its own values of reliability and also "factors of effect" for a number of error producing conditions (Embrey and Kirwan 1983).

The basis for applying HEART is a classification of tasks into the generic types, enabling calculation of the proposed nominal human unreliability for the execution of the tasks. The method consists of the following steps (Kirwan 1994).

(1) Refine the task in terms of its generic, proposed and nominal level of human unreliability.
(2) Identify the full range of sub-tasks that a system operator would be required to complete within a given task.
(3) Determine a nominal human reliability for the particular task, usually by consulting local experts. Establish a 5–95th percentile confidence range based around this calculated score.
(4) Identify EPCs, which are evident in the given situation and would have a negative effect on the outcome.

Table 6.19 Example of HEART calculation format (Kim et al. 2006b)

External error mode	Generic task type	GEP	EPC (max, rating)	EPC value	HEP (GEP × EPC)
Driver fails to check the signal	Routine, highly practiced, rapid task involving a relatively low level of skill	0.02	Ability to detect ad perceive (10, 0.5)	$PF_1 = (9 \times 0.5) + 1 = 5.5$	0.286
			Unfamiliarity (17, 0.1)	$PF_2 = (16 \times 0.1) + 1 = 2.6$	
Driver checks a wrong signal	Routine, highly practiced, rapid task involving a relatively low level of skill	0.02	Ability to detect ad perceive (10, 0.5)	$PF_1 = (9 \times 0.5) + 1 = 5.5$	0.286
			Unfamiliarity (17, 0.1)	$PF_2 = (16 \times 0.1) + 1 = 2.6$	

(5) Calculate a final estimate of the human error probability (HEP).

The HEP calculation is given by the following equation

$$\text{Final HEP} = \text{NEP} \times \prod [R(i) \times (W(i) - 1) + 1] \tag{6.25}$$

where,

HEP = the human error probability

NEP = the nominal error probability given for a selected generic task type

$R(i)$ = the rating of the ith EPC

$W(i)$ = the weighting of the ith EPC

HEART has found increasing usage in the railway sector, and an example of its use is given below (Kim et al. 2006b). A type of event known as signal passed at danger (SPAD) is the most frequent contributor to railway accidents. Causes of SPADs include:

- Signal not seen due to bad visibility.
- Misjudgement of which signal applies to the train in question.
- Misunderstanding or disregard of the signal.

Table 6.19 contains the quantification of the above causes in HEART.

In order to mitigate the HEP, HEART goes further than other techniques and suggests error reduction approaches for each one of the EPCs (Kirwan 1994). For instance, better training and/or communication procedures can reduce the detection problems of signal checking.

HEART is one of the few HRA techniques to have been independently empirically validated (Bell and Holroyd 2009). Based upon this, a significant correlation between the true and the assessed values of HEP has been determined. Currently UK's Railways Safety Standards Board (RSSB) is developing an HRA model based on HEART (Gibson and Kirwan 2008b).

6.7.3 Aviation

It has been stated that between 70 and 80% of aviation accidents are attributed, at least in part, to human errors (Wreathall et al. 2003). Hence human reliability analysis has a major role to play in risk analysis and the prevention of accidents in the future. Aviation is itself a broad term, with a large variety of operations and technology and many elements inter-related in a complex manner. Air traffic control lends itself well to HRA techniques, especially as it involves the complex interaction of humans, equipment and procedures, and this has led to the development of dedicated HRA techniques for the domain. This section outlines below three techniques used in air traffic control.

6.7.3.1 THERP

Technique for human error rate and prediction (THERP) is probably the best known first generation HRA method (Eurocontrol 2007). As Swain and Guttmann (1983) note the aim of THERP is to calculate the probability of the successful performance of the activities necessary for the accomplishment of a task. These calculations are based on predefined error rates known also in this technique as human error probabilities (HEPs), and success is defined as the complement to the probability of making an error. THERP involves performing a task analysis to provide a description of the performance characteristics of the human tasks being analyzed. The results of the task analysis are represented graphically in a so-called HRA event tree that is a formal representation of the required sequence of actions. The nominal probability estimates from the analysis of the HRA event tree are modified for the effects of the sequence specific performance shaping factors (PSFs), which are factors that considerably affect the practicality of an "action and influence HEP, and may contain factors such as the dependence between and within operators, stress levels, experience, the quality of information provided, training, fatigue" (Kyriakidis 2009a, b).

THERP is summarized is six main steps (Eurocontrol 2007):

(1) Define the system failures that may be influenced by human errors and for which probabilities are to be estimated.
(2) Identify, list, and analyze the human operations performed and their relationships to system tasks and function of interest, i.e. undertake a task analysis.
(3) Estimate the relevant HEPs, i.e. predicted error rates.
(4) Determine the effects of human errors on the system failure events of interest.

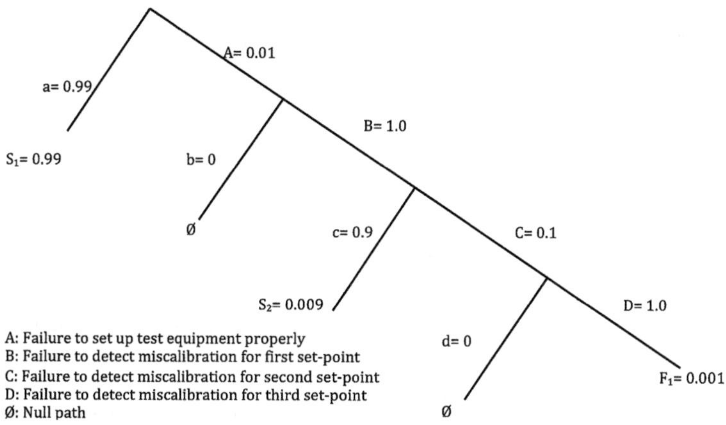

Fig. 6.35 Example of human reliability analysis event tree (Kirwan 1994)

(5) Recommend changes to the system in order to reduce the system failure rate to an acceptable level.
(6) Review the consequences of proposed changes with respect to availability, reliability, and cost benefit.

The THERP procedure can be outlined as follows (Kirwan 1994):

Phase 1: familiarisation

- Plant visit
- Review information from system analysis

Phase 2: qualitative assessment

- Talk or walk-through
- Task analysis
- Develop HRA event trees

Phase 3: quantitative assessment

- Assign nominal HEPs
- Estimate the relative effects of PSFs
- Assess dependence
- Determine success and failure probabilities
- Determine the effects of recovery factors

Phase 4: incorporation

- Perform a sensitivity analysis, if warranted
- Supply information to system analysis

In THERP an event tree is used for HEPs modeling. Event trees represent a binary decision process, i.e. success or failure in task performance as the only

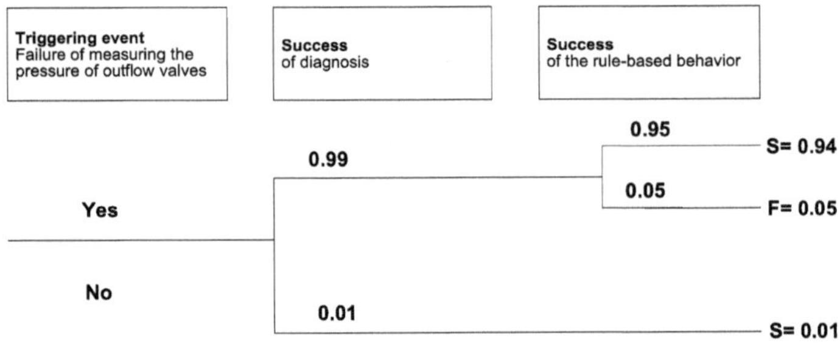

Fig. 6.36 Example of HEP estimation with THERP method (based on ETH-LSA)

Table 6.20 Example of THERP calculation format

Probabilities calculation by using THERP	
Probability of false diagnosis	$P_{(F)\ 50} = 0.01 \rightarrow P_{(S)\ 50} = 0.99$
Probability of false behavior	$P_{(F)\ 50} = 0.05$ à $P_{(S)\ 50} = 0.95$
Success paths	$S = 0.99 \times 0.95 \approx 0.94$
Failure paths	$F_1 = 0.99 \times 0.05 \approx 0.05$
	$F_2 = 0.01$
Sum of failure paths probabilities	$\Sigma\ P_{(F)\ 50} = 0.06$
Probability of system to fail	Pr(system failure) = Pr(triggering event) $\times\ \Sigma\ P_{(F)\ 50}$

possibilities (Fig. 6.35). The success and failure probability outcomes sum to unity. A great advantage of using the event tree is that it is possible to explicitly depict paths by which errors can be recovered.

An example of THERP implementation to estimate the HEP is outlined as follows. A civilian depot technician is responsible for gauging the pressure of outflow valves of a KC-135 aircraft, which is being pressurised at ground level. To assess the human action, the "diagnosis" and the "behavior" should be combined. The following assumptions should be considered:

- The outflow valves were capped off during a 5-year overhaul and never re-opened.
- A civilian depot technician was using a homemade gauge, and not the standardized procedure.
- The triggering event is described as the failure of the technician to measure the pressure of outflow valves.
- The probabilities of success or failure reaction are hypothetical.

Figure 6.36 depicts the corresponding event tree. Using THERP, human reliability and the operational technician's assessment of performance could look like Table 6.20. THERP has been found to achieve a reasonable level of accuracy (Bell and Holroyd 2009). It was developed for the probabilistic risk assessment (PRA) of nuclear power plants but has also been applied to other sectors such as oil and gas offshore and healthcare (Bell and Holroyd 2009).

In summary, THERP has been described (Kirwan 1994) as one of the few complete first generation HRA techniques, in the sense that it describes both how events should be modeled (event trees) and how they should be quantified. The dominance of the HRA event tree, however, means that the classification scheme and the model are necessarily limited, as the event tree can only account for binary choices (success–failure). It is thus difficult to introduce more complex error modes in THERP.

6.7.3.2 ATHEANA

A technique for human event analysis, or ATHEANA, is a second generation HRA method developed for the US Nuclear Regulatory Commission (Forester et al. 2007), designed to support the understanding and quantification of human failure events[8] (HFEs) in nuclear power plants. However the approach is suitable for application in other industries (Bell and Holroyd 2009). ATHEANA's basic idea claims that significant human errors are a result of "error-forcing contexts" (EFCs), which are defined as combinations of plant conditions and other influences that make an operator error.

ATHEANA is an HRA methodology designed to search for such EFCs, by using and integrating knowledge and experience in engineering, PRA, human factors, and psychology with plant-specific information and insights from the analysis of serious accidents (Forester et al. 2007).

ATHEANA can be summarized in nine step accidents (Forester et al. 2007):

(1) Define and interpret the issue (in this step analysts define the objective that is to be achieved by performing the HRA).
(2) Define the scope of the analysis.
(3) Describe the PRA accident scenario and its nominal context.
(4) Define the corresponding HFE or unsafe actions[9] (UA), which may affect the task in question.
(5) Assess information relevant to human performance and characterize the factors that could lead to potential vulnerabilities.
(6) Search for plausible deviations from the PRA scenario.
(7) Evaluate the potential for recovery.
(8) Estimate the HEPs for the HFEs/UAs.
(9) Incorporate each HFE/UA and corresponding HEP into the PRA.

---

[8] An HFE is a basic event modeled in the logic models of a PRA (logic) and represents a failure of a function, system, or component that is the result of one or more unsafe actions. A human failure event reflects the PRA system's modeling perspective (Kirwan 1994).

[9] An *unsafe action* is an action inappropriately taken, or not taken when needed, by plant personnel that results in a degraded plant safety condition (Kirwan 1994).

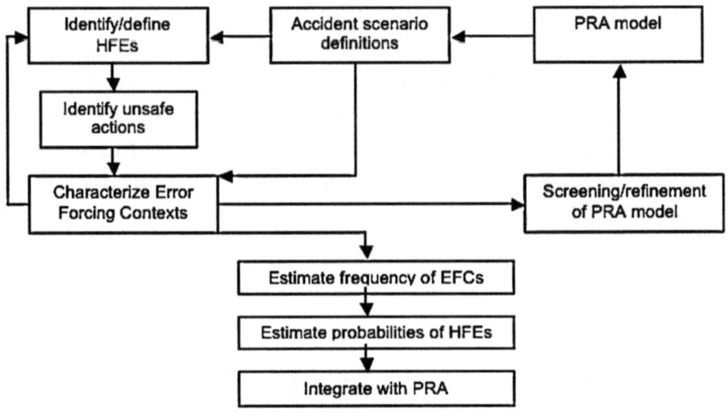

Fig. 6.37 ATHEANA method (Hollangel 1998)

Figure 6.37 shows the flow diagram for the application of ATHEANA. ATHEANA contains two important loops. The first is from the characterization of the EFCs to the identification of the HFEs. This recognizes that an improved description of the context may enable a better identification of HFEs and this in turn may amend the description of the context. The second involves the characterization of the EFCs into the PRA model. This suggests that the outcome of the qualitative part of HRA may be used to modify the underlying PRA model, for instance by highlighting conditions or human-system interactions that have been missed in the first place.

For a given scenario, *S*, the final quantification step is expressed by:

$$P\left(\frac{E}{S}\right) = \sum_{\substack{\text{unsafe}\\ \text{action}(i)}} \sum_{\substack{\text{error}\\ \text{forcing}\\ \text{context}(j)}} Pij(S) \tag{6.26}$$

- *P*(*E*/*S*) is the probability of the HFE in the scenario *S*
- *Pij*(*S*) is the probability of the unsafety action *i* resulting from EFC*j* in scenario *S*

A peer review of ATHEANA, its documentation, and the results of an initial test of the method were held over a 2-day period in 1998 (Bell and Holroyd 2009). The reviewers' general opinion of ATHEANA was that the method represents a significant improvement in HRA methodology; it is a useful and usable method; and it is a "good alternative to first-generation HRA approaches" (Bell and Holroyd 2009). However, the reviewers also note that the method for quantification is weak, and that the quantitative results are excessively dependent on expert judgement, hence possibly it has low credibility as a method and needed to be improved and extended (Forester et al. 1998).

Whilst this example has been generic, literature indicates that ATHEANA can be applied to air traffic control domain (Subotic 2007).

6.7.3.3 CARA

Building on the basic quantification framework of HEART, the controller action reliability assessment (CARA) is a HRA technique, used to quantify human performance in the context of air traffic management (ATM) safety assessments. There are at least four clear application areas for HRA (Kirwan and Gibson 2007):

(1) Individual concept element safety cases e.g. a safety case for a new conflict resolution system, or for an arrival manager.
(2) Unit safety cases e.g. a safety case for Maastricht upper airspace centre, or another air traffic control centre or airport.
(3) A human factors-driven HRA focusing on a specific current problem area or a proposed change that may have impact on human error and recovery performance.
(4) System-wide safety cases, for next generation ATM systems e.g. in Europe for SESAR, or in the US potentially for Next Gen.

Experience indicates that the most successful HRA approaches have been flexible and tailored to specific industries. Such tools are useful for most safety case needs; it appears sensible, therefore, that ATM developed a similar approach, using generic task types relevant to the industry and safety case needs i.e. typical tasks or errors modeled in safety cases, with the appropriate needs of modification factors e.g. related to traffic, weather, human machine interface (HMI) (Kirwan and Gibson 2007). HEART was selected as the basis for CARA's development because it has been the subject of validation exercise (Gibson and Kirwan 2008a, b) and also was already applied to different domains such as the railway (Kim et al. 2006b) and nuclear industries (Gibson and Kirwan 2008a, b).

CARA introduces the following three main key elements:

- *Generic task types (GTTs)*: During an HRA, analysts will be asked to quantify specific tasks. The GTT selected is the one that best matches the specific task being assessed; it is associated with a human error probability and therefore this provides an initial quantification for the task being assessed. A set of GTTs, which are specific to the ATM environment and have been quantified, using human performance data, has been developed for CARA.
- *Error producing conditions (EPCs)*: EPCs are factors, which are predicted to influence human performance negatively and therefore increase the generic human error probability associated with a GTT. EPCs can be amongst others, 'time pressure' or 'operator inexperience'. The technique also defines a numerical value, called the 'maximum affect', which reflects the maximum impact of an EPC on a task.
- Numerical values have been developed for EPCs and maximum affects in CARA.
- *Calculation method*: For ATM, CARA uses HEART's calculation method to estimate the HEP and the strength of affect of EPCs through a weighting process.

HEP calculation is given from (Gibson and Kirwan 2008a):

Table 6.21 CARA generic task types (Gibson and Kirwan 2008a)

Task context	Generic task type	HEP	Uncertainty bounds
A. Offline tasks	A. Offline tasks	0.03	–
B. Checking	B1. Active search of radar or FPS, assuming some confusable information on display	0.005	0.002–0.02
	B2. Respond to visual change in display (e.g. aircraft highlighted changes to low-lighted)	0.13	0.05–0.3
	B.3 Respond to unique and trusted audible and visual indication	0.0004	–

Table 6.22 Example of proposed CARA error-producing conditions (Gibson and Kirwan 2008a)

HERA element	CARA error-producing conditions	Maximum affect
Training and experience	Unfamiliarity and adequacy of training/experience	20
Environment	Environment-controller workplace noise/lighting issues, cockpit smoke	8
Personal factor issues	High emotional stress and effects of ill health	5

$$\text{HEP} = \text{GTT} \times [(\text{EPC}_1 - 1) \times \text{APOA}_1 + 1] \times \ldots \times [(\text{EPC}_n - 1) \times \text{APOA}_n + 1] \quad (6.27)$$

where:

GTT = the human error probability associated with a GTT

EPC = the maximum affect associated with an EPC

APOA = is the assessed proportion of affect value between 0.05 and 1, where 0.05 is a very weak effect and 1 is a full affect.

Tables 6.21 and 6.22 depict some of the GTTs and proposed EPCs selected for CARA in order to comply with air traffic management needs (Gibson and Kirwan 2008a).

CARA has been applied to a safety scenario related to an aircraft landing guidance system and the results compared with those from HEART for the same scenario (Gibson and Kirwan 2008a). This study highlighted a number of findings related to the difference in applying CARA and HEART. The HEART application used only two generic task types, whereas the CARA application used six different generic type tasks descriptions; CARA's GTTs were better tailored to the specifics of the scenario and the maintenance tasks. In general, the choice of GTT was quite simple and consequently fewer EPCs were required for the application of CARA.

The calculated values for the scenario using CARA were generally within one order of magnitude to those corresponding to the HEART calculated values (Fig. 6.38). It should be noted that the results are not a reflection on the reliability of either HEART or CARA, rather they merely compare the quantification of

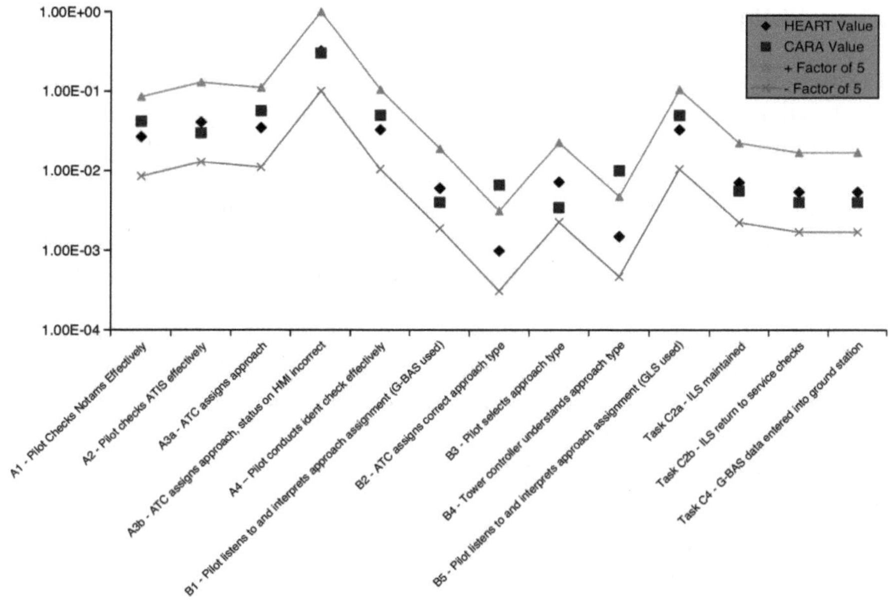

Fig. 6.38 CARA and HEART human error probabilities for the aircraft landing guidance system scenario (Gibson and Kirwan 2008a)

outcomes if the CARA approach is applied, with those calculated using HEART. While not a reliability study, this is at least a positive indication of convergence between the two techniques.

Although the process of validating CARA is ongoing, initial results indicate that it has been successfully applied to three safety cases, and provides an initial indication that HRA can be used to deal with human factors arguments in a quantified ATM safety case context (Gibson and Kirwan 2008a).

Due to the fact that CARA is derived from HEART, it should be noted in that both methods use exactly the same structure, although they have different GTTs, EPCs and maximum effect multipliers for the EPCs. This being the case, reader should be aware that Eqs. 6.25 and 6.27 are identical.

6.7.4 Road Transport of Dangerous Goods

The domain of road transport does not seem to have developed HRA techniques when compared to railways and aviation, although undoubtedly human error has a major part to play in road accidents. One application that has seen the development of HRA techniques is that of the transport of dangerous goods, as outlined below.

6.7.4.1 SLIM

The success likelihood index method (SLIM), a first generation HRA technique developed by Embrey et al. (1984) gives a structured judgement about error probabilities in both procedural and cognitive tasks (Eurocontrol 2007) and is used for the purposes of evaluating the probability of a human error occurring throughout the completion of a specific task. SLIM is a decision-analytic approach to HRA that uses expert judgement to quantify performance shaping factors (PSFs). Such factors are used to derive a success likelihood index (SLI) (Robles et al. 2008) which represents the overall belief of experts and analysts, regarding the positive or negative effects of the PSFs on the likelihood of success for the task under consideration. The SLI is calibrated against existing data to derive a final HEP. Experts choose the PSF's, which are considered as the most significant in relation to the context in question.

SLIM methodology is described in nine steps:

(1) Selection of the expert panel
(2) Definition of situations and subsets
(3) Elicitation of PSFs
(4) Rating of the tasks on the PSF scale
(5) Ideal point elicitation and scaling calculations
(6) Independence checks
(7) Weighting procedure
(8) Calculation of SLIs
(9) Conversion of SLIs into probabilities

Typical PSFs used in SLIM include; time pressure or stress levels; the quality of information or quality of interface; the quality of procedures; the task complexity; the consequences as perceived by the operator; the required amount of teamwork and the adequacy of training or level of competence (Embrey et al. 1984).

The SLI for n PSFs is deduced using the following equation:

$$\mathrm{SLI} = \sum_{i=1}^{n} r_i \times w_i \tag{6.28}$$

where

- w_i the quality weighting
- r_i the rating factor

The SLIs are converted into probabilities by using Eq. 6.29

$$\log(P) = a\ \mathrm{SLI} + b \tag{6.29}$$

where, HEP the success probability and a, b are constants.

The constants a (*slope of the line*) and b (*the intercept of the line with the vertical axis*) can be derived by processing simultaneous equations, as long as at

Table 6.23 Example of performance shaping factors rating (Kirwan 1994)

Errors	Training	Procedures	Feedback	Perceived risk	Time
V0101 open	6	5	2	9	6
Alarm mis-set	5	3	2	7	4
Alarm ignored	4	5	7	7	2

least two calibration probabilities have been assessed within each task subset or from already estimated HEPs given by the THERP handbook data and the data of Kirwan (1994).

An example of SLIM application can be seen from the transport of dangerous goods, in particular the task of de-coupling a filling hose from a chemical road tanker (Kirwan 1994).

The closure of a valve located upstream of the filling hose, known as V0101, by the operator is a crucial part of the procedure and he/she may forget this. The human error of interest in this situation is "Failure to close V0101 prior to decoupling filling hose". In this case, the decoupling operation is simple and discrete, hence the failure occurs catastrophically rather than in a staged fashion.

The "expert panel" required to carry out the HRA may consist of two operators possessing approximately 10 years of experience, a human factors analyst and a reliability analyst who is familiar with the system and possesses a degree of operational experience. The panel is requested to determine a set of PSFs, which are applicable to the task, in question within the context of the overall system (Kirwan 1994). Having identified the PSFs, the panel proposes the most important ones in the specific scenario. For this example, in this situation, the panel may identify the following major PSFs as affecting human performance: training, procedures, feedback, perceived level of risk, and time pressure.

Assessing the situation within the context of the task under assessment, the panel is asked to provide further possible human errors that may occur and have the potential of affecting performance (in this case mis-setting or ignoring an alarm are selected).

For each of these, the experts are required to establish the degree to which each is either optimal or sub-optimal for the task under assessment, based on a scale of 1–9, with the latter being the optimal rating. For three human errors that have been identified, the ratings decided for each are provided in Table 6.23.

If each factor is of equal importance, it is then possible to obtain the summation of each row of ratings and come to the conclusion that the row with the lowest total rating is most likely to occur, in this case, it would be alarm mis-set.

However, usually the experts are in agreement that the PSFs given above are not of equal weighting. Hence, in this scenario the experts decided that perceived risk and feedback are deemed to be of the greatest importance, twice as much as training and procedures, and these latter two are considered to be one and a half times more important than the factor of time. The time factor is of considered of minimal importance in this case as the task is routine and is therefore, not time-limited. Table 6.24 indicates these weightings.

Table 6.24 Example of performance shaping factors weighting (Kirwan 1994)

PSF	Importance	PSF	Importance
Perceived risk	0.3	Procedures	0.15
Feedback	0.3	Time pressure	0.10
Training	0.15	Sum	1.0

Table 6.25 Success likelihood index (total) calculation (Kirwan 1994)

Weighting	PSFs	V0101 open	Alarm mis-set	Alarm ignored
0.3	Perceived risk	$0.3 \times 9 = 2.7$	2.10	2.10
0.3	Feedback	0.60	0.60	2.10
0.15	Training	0.90	0.75	0.60
0.15	Procedures	0.75	0.45	0.75
0.10	Time pressure	0.60	0.40	0.20
	SLI (total)	5.55	4.30	5.75

Using the figures for the scaled rating of the PSFs and weighting their importance, the SLIs can be calculated for each part of the task under assessment, Table 6.25. The results indicate that as the SLI for 'alarm mis-set' is the lowest, this is the most probable error to occur throughout the completion of the task.

In order to transform SLIs into HEPs, it is first necessary to "calibrate" the SLI values. Assuming that two additional tasks A and B have been assessed, with HEP values of 0.5 and 1×10^{-4} and SLIs of 4.00 and 6.00, respectively, and based on Eq. 6.29 $a = -1.85$ and $b = 7.1$, then the final HEP for V0101 is estimated as:

$$\log(P) = a \times \text{SLI}(\text{V0101 open}) + b = -1.85 \times (5.55) + 7.1 = -3.1675 \Rightarrow$$
$$\Rightarrow P = 10^{-3.1675} = 0.00068 \tag{6.30}$$

The corresponding HEPs for "alarm mis-set" and "alarm ignored" are calculated likewise. The level of accuracy associated with SLIM is indeterminate due to lack of data. However, its theoretical validity is at a reasonably high level. (Embrey and Kirwan 1983) carried out a validation of SLIM's expert judgment which showed that a further development and an improvement in calibration process is needed. SLIM has been applied in the nuclear and chemical industries (Bell and Holroyd 2009).

It should be mentioned that SLIM is used as an interactive computer program called multi-attribute utility decomposition (MAUD). The developers of the technique strongly recommend that SLIM be implemented using the software and have termed the overall approach SLIM-MAUD. However, the software has not been updated for application with current computer technology (Forester et al. 2006).

Lastly, it should be noted that there are more than one variant of SLIM, such as the failure likelihood index methodology (FLIM) which has had extensively been applied in several countries (Chien et al. 1988).

6.7.5 Electrical Network

The electric power system is a complex, large-scale and vulnerable infrastructure. It qualifies as critical because a reliable power supply is essential for many services, social and economic activities as well as for the functioning of other vital infrastructures. The introduction of market liberalization has substantially complicated the situation since the network is now asked to transport power in ways that it was not originally designed to do. In conjunction operators deal with situations, responsibilities, and duties that may have not been addressed before. Hence, the system's efficient functioning and reliability are related not only to equipment's adequacy but also to the operators' performance and coordination. Therefore, it is essential to investigate human influence for a deeper and more comprehensive analysis (Kyriakidis 2009b). The cognitive reliability and error analysis method (CREAM) is a HRA technique that has already been applied to the analysis and estimation of human error probability in the electricity domain.

6.7.5.1 CREAM

The cognitive reliability and error analysis method (CREAM) is a second generation HRA and enables an analyst to:

Identify those parts of the work, as tasks or actions, that require or depend on human cognition, and which therefore, may be affected by variations in cognitive reliability.
Determine the conditions under which the reliability of cognition may be reduced, and therefore, where these tasks or actions may constitute a source of risk.
Provide an appraisal of the consequences of human performance on system safety which can be used in a probabilistic safety analysis (PSA).
Develop and specify modifications that improve these conditions and hence serve to increase the reliability of cognition and reduce the risk. (Hollangel 1998)

The first three steps are the core of CREAM, while the last aims to ensure that proper conclusions are drawn from the analysis, and that the necessary changes to the system are correctly specified.

CREAM provides the core functionality of these services, i.e. the concepts, the classification system, the cognitive models, and the techniques. In order to be properly used, it is necessary to supplement with application or plant specific information e.g. in the form of values for specific performance parameters, the detailed operational and process knowledge that define the context (Hollangel 1998).

The model is based on a fundamental distinction between competence and control. A classification scheme consists of a number of groups that described the phenotypes (effects-error modes) and the genotypes (causes) of the erroneous actions. Phenotypes refer to what is observable in the given system, while genotypes are the categories that can be used to describe that which can bring about the

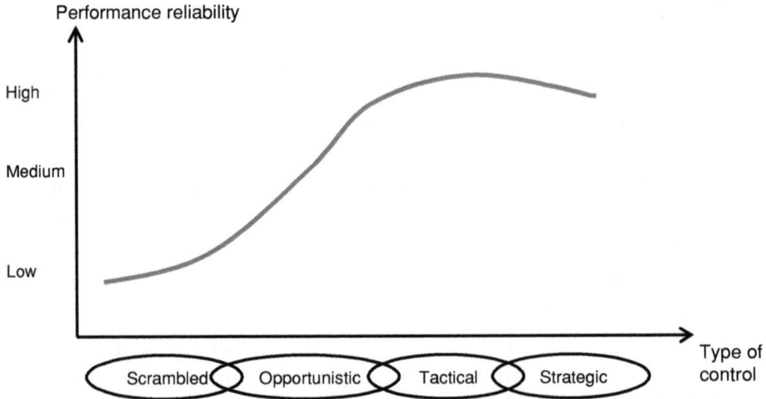

Fig. 6.39 Proposed relation between control mode and reliability (Hollangel 1998)

effects. The CREAM classification scheme is not only used by the analysts to predict and describe how errors could potentially occur, but also allows them to define the links between the causes and consequences of the error under analysis.

Within the CREAM classification scheme there are three categories of causes; person-related genotypes; technology-related genotypes and organisation-related genotypes. In CREAM the following four control modes are suggested (Hollangel 1998):

- *Scrambled control*: choice of the next haphazard action, little or no thinking involved, task demands high, loss of situational awareness, momentary panic.
- *Opportunistic control*: choice of action based on present conditions, little planning or anticipation, unclear context, constraint time.
- *Tactical control*: performance based on planning, follows procedures/rules.
- *Strategic control*: person considers the global context, wider time horizon, robust performance. The functional dependencies between task steps are important.

According to Hollangel (1998), when the level of an operator's control rises, so too does his/her performance reliability, as illustrated in Fig. 6.39. CREAM can be used in several different ways as (Hollangel 1998):

- A stand-alone analysis method, for either retrospective or prospective analyses, using a consistent taxonomy for error modes and error causes.
- Part of a larger design method for complex, interactive systems.
- A HRA in the context of an integrated safety analysis or (PRA).

CREAM approaches the quantification part in two steps by providing a basic and an extended method (Hollangel 1998). Based on this, a list of operator activities is produced from which a common performance condition (CPC) analysis is carried out. There are nine CPCs: adequacy of organisation; working conditions; adequacy of the man-machine interface and operational support; availability of procedures/plans; number of simultaneous goals; available time; time of day; adequacy of training and experience and the quality of crew collaboration.

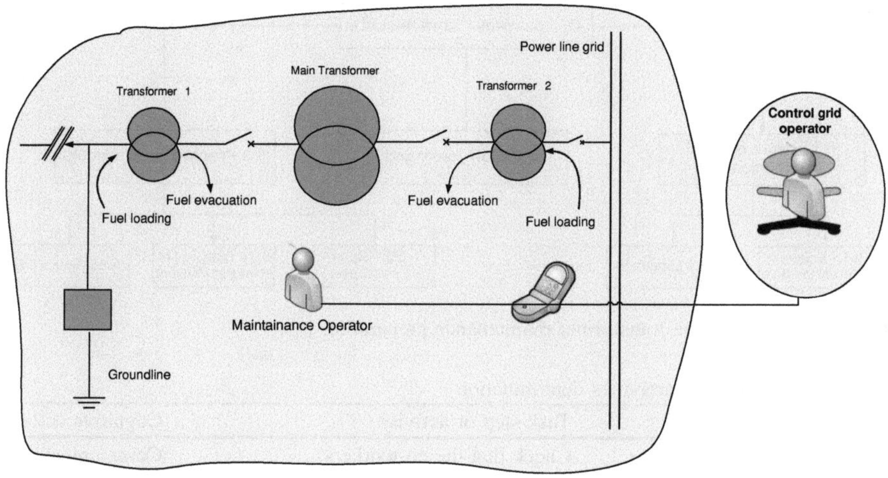

Fig. 6.40 Network's maintenance procedure

For each activity a CPC level is determined, for example adequacy of training and experience is described as high experience, low experience or inadequate. The expected effects of these levels of experience on performance are, respectively improved, not significant and reduced. The method outlines a means of quantifying these descriptors. The sum of the performance reliability (i.e. improved, not significant and reduced) for each CPC gives a combined CPC score e.g. for the nine CPCs the result may be [9, 0, 0], which would be the least desirable situation as all CPCs indicate reduced performance reliability, whereas [0, 2, 7] describes a much more desirable situation.

The following example describes the extended CREAM method applied to an electrical power plant (Kyriakidis 2009b). The working scenario describes a maintenance procedure that very often takes place in an electrical circuit as depicted in Fig. 6.40, involving two operators: an internal "Network Control" operator and an external control grid operator. The two operators communicate by telephone during the whole procedure. An official report that contains details about the purpose, the steps, and the employees responsible for the tasks should be filled out at the beginning of the procedure.

The duties of the "network control" operator during the process are:

- Cut off the power from the circuit
- Communicate with the external operator and check the sequence of the working steps
- Re-connect the power to the circuit

The external operator should:

- Open the switches that connect the transformers among them and to the grid
- Disconnect the main transformer
- Disconnect the transformers 1 and 2

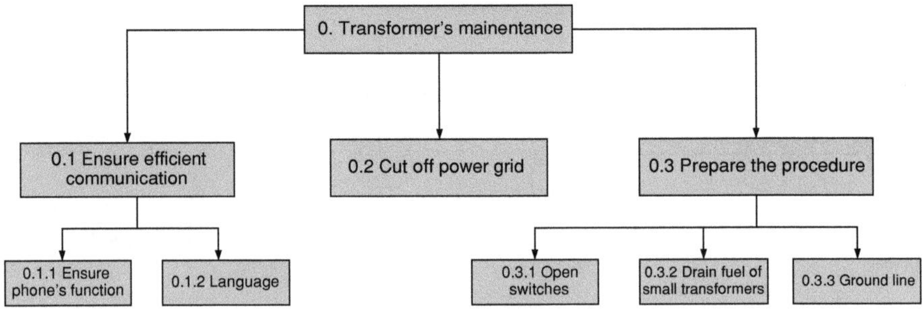

Fig. 6.41 HTA for the transformer maintenance preparation

Table 6.26 Cognitive activities determination

Step	Goal	Task step or activity	Cognitive activity
0.1.2	Language	Check that the co-workers can understand each other	Communicate
0.2	Cut off grid power	Ensure that the main grid power is off	Verify
0.3.1	Open switches	Ensure that all switches are open	Execute

- Evacuate the fuel from the transformers' tanks. Hence, even if by mistake a switch is not off anymore, the transformer cannot function
- Connect the ground line to the ground land
- Maintain the transformer
- Reload the fuel to transformers' tank
- Close the switches
- Inform operators to reconnect the power to the grid

Figure 6.41 illustrates the hierarchical task analysis (HTA) before the start of the transformer's maintenance. The following paragraphs describe step by step the estimation of the failure probabilities.

Step 1: In the first step cognitive activities for each one of the tasks have been defined based on the HTA, as depicted in Table 6.26.

Step 2: Having determined the cognitive activities, the next step is to define the corresponding cognitive functions (Table 6.27) by matching the cognitive activities with the cognitive demand matrix (cognitive failure types).

Step 3: The third step is related to the identification of the most likely cognitive function failures. To accomplish this step the analyst can build an appropriate table, e.g. Table 6.28. In this table, the dominant cognitive failure activity for each one of the tasks is defined.

For example, coordination cognitive activity involves the cognitive functions of planning and execution. In this case the dominant function is the execution. Furthermore, in this table the type of the generic failure type that describes the case is determined.

Step 4: The determination of the cognitive failure probability (CFP) is the next step of this procedure. When HRA is performed as a part of a PSA, the event tree

Table 6.27 Cognitive demands

Step	Task step or activity	Cognitive activity	Obs	Int	Plan	Exe
0.1.2	Check that the co-workers can understand each other.	Communicate				•
0.2	Ensure that the main grid power is off.	Verify	•	•		
0.3.1	Ensure that all switches are open.	Execute				•

Obs observation, *Int* interpretation, *Plan* planning, *Exe* execution

Table 6.28 Dominant failure activities table

Step	Cognitive activity	Obs			Int			Plan		Exe				
		O1	O2	O3	I1	I2	I3	P1	P2	E1	E2	E3	E4	E5
0.1.2	Communicate													•
0.2	Verify			•										
0.3.1	Execute											•		

Table 6.29 Assessment of the effects of CPC on the procedure

PSF name	Condition	0.1.2 E5	0.2 O3	0.3.1 E4
Working conditions	Compatible	1.0	1.0	1.0
Organization culture	Efficient	1.0	1.0	1.0
Man machine interface	Supportive	1.0	0.5	1.0
Stress—available time	Temporary inadequate	1.0	1.0	1.0
Fatigue	Tired	1.2	1.2	1.2
Training and experience	Adequate, low experience	1.0	1.0	1.0
Team collaboration	Efficient	1.0	1.0	1.0
Operational procedures	Appropriate	0.8	0.8	0.8
Task complexity (number of goals)	More than capacity	2.0	2.0	2.0
Total weighting factor		1.92	0.96	1.92

(or fault tree) defines the activities for which a failure probability must be calculated. These activities will typically only represent a subset of the steps in the event sequence. The final adjusted CFP is a value that is derived by multiplying the nominal CFP (Hollangel 1998) with a weighting factor obtained from an influence to the task, as shown in the Tables 6.29 and 6.30. The total influence of CPCs for each cognitive failure is found by multiplying all the CPCs.

Step 5: From the calculations, the failure probability with the highest value is related to possible language communication problems between operators at 0.0576 6×10^{-2}.

Based on the method's basic diagram, the probability value is located either in the interval that is described as tactical or in the interval described as opportunistic, as shown in Fig. 6.42.

Table 6.30 Adjusted cognitive failure probabilities for the procedure

Step	Task step or activity	Error mode	Nominal CFP	Weighting factor	Adjusted CFP
0.1.2	Check that the co-workers can understand each other.	E5	3.0 E-2	1.92	5.76 E-2
0.2	Ensure that the main grid power is off.	O3	2.0 E-2	0.96	1.92 E-2
0.3.1	Ensure that all switches are open.	E4	1.0 E-3	1.92	1.92 E-3

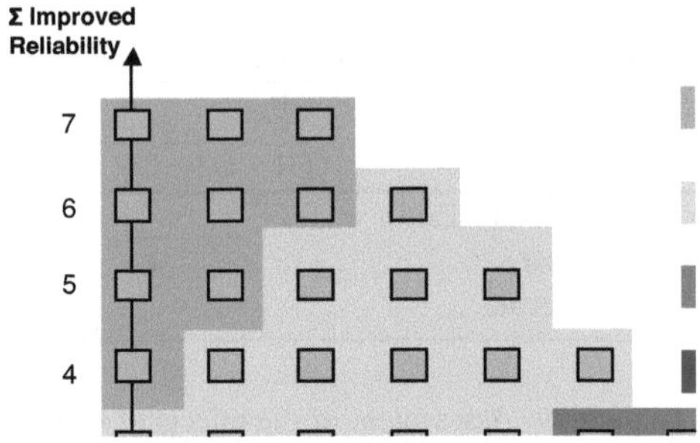

Fig. 6.42 Relations between common performance conditions score and control modes (Hollangel 1998)

However, due to the fact that CREAM illustrates a sense of reliability, this method mainly provides a general reliability perception and not a specific probability value. Nevertheless, it provides analysts with a sufficient perception of the safety and reliability levels. In addition, it is an easily applicable method for several scenarios and it can be applied in a wide variety of domains. Based on the findings of the previous example:

- CREAM can be applied in the area of electricity infrastructure as it can provide analysts with a sense of safety and reliability regarding the human contribution to the procedure.
- As the technical systems and equipment in electricity infrastructure are continuously improved, human reliability should be also ensured and enhanced.
- CREAM addresses a significant weakness, related to analysts' personal training and experience known as "expert judgment". Therefore, it is necessary to ensure that the analysts who conduct the safety researches have similar educational and training background. Otherwise, the danger of different findings even with the same database cannot be easily avoided.

The process of assessing the validity and reliability of CREAM is still ongoing (Bell and Holroyd 2009).

6.7.6 Public Health

In healthcare, a non-physical engineered CI, a "medical error" can be defined as *an inaccurate or incomplete diagnosis and/or treatment of a disease; injury;*

infection or any other ailment, which can compromise patients' safety. Mortality and morbidity resulting from medical errors compel a better and deeper understanding of health care as a system (Wreathall and Nemeth 2004).

Given the large number of HRA techniques applied to other domains, none has been designed for the healthcare industry exclusively. In recent years, an interest in applying HRA techniques to the healthcare industry has grown.

The existing HRA techniques, which could be applied to healthcare, can be grouped into five main categories (Lyons et al. 2004):

- Data collection (collection of information on incidents, goals, tasks is taking place)
- Task description (taking the collected data and portraying these in useful form)
- Task simulation (simulating the task as described and changing aspects to identify problems)
- Human error identification and analysis (uses task description, simulation and/or contextual factors to identify the potential error)
- Human error quantification (estimated the probability of the identified errors)

Techniques may be used either separately or in combination and they have been grouped according to their principal types and the purpose of the analysis.

6.7.6.1 Fault Trees

Even though not a HRA technique, fault trees are used to analyze and quantify human error probabilities in healthcare. Fault tree diagrams represent cause and effect relations among events that culminate in a "top event". Logic symbols at each intersection indicate what is required to occur for the condition to be satisfied (Wreathall and Nemeth 2004). An example of a fault tree that applied to a medication error is illustrated in Fig. 6.43 (Lyons et al. 2004): Fault trees allow analysts to quantify the error probability and also allow them to analyze the task and explore the possible scenarios that may lead to top event.

6.7.6.2 SHERPA

In addition to fault trees, several HRA techniques can be applied to healthcare, such as the systematic human error reduction and prediction approach (SHERPA). SHERPA is a first generation technique developed by Embrey in 1986 as a human-error prediction technique that also analyzes tasks and identifies potential solutions to errors in a structural manner. The technique was initially designed to be applied to the process industries, such as nuclear power plants; oil and gas extraction; petrochemical processing (Stanton et al. 2005) and in recent years also to the healthcare industry. A study conducted in 2006 applies SHERPA to a task of administering drugs to hospital patients (Lane et al. 2006).

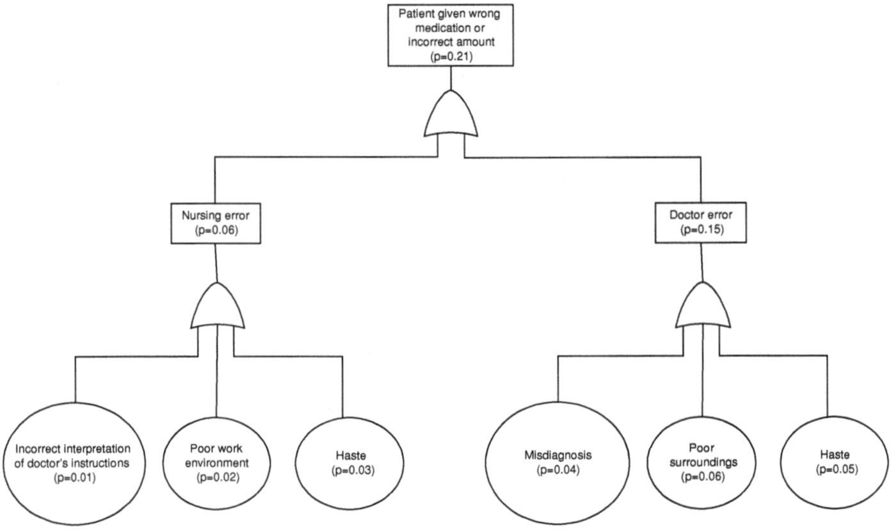

Fig. 6.43 A fault tree applied to medication error (Lyons et al. 2004)

SHERPA contains eight steps (Stanton et al. 2005):

1. Hierarchical task analysis
2. Ordinal probability analysis
3. Criticality analysis
4. Remedy analysis
5. Recovery analysis
6. Task classification
7. Human-error identification
8. Consequence analysis

The example below illustrates how SHERPA can be implemented in the healthcare industry and the full description of the example is given in the literature Lane et al. (2006).

A study was conducted about medication to hospital patients. For each medication the analyst was asked what would happen when someone mistakes a particular drug package for something else; uses the wrong amount of the drug; gives the drug to the wrong patient; gives the drug by the wrong route; gives the wrong rate of a drug and so on. The analyst then continued to consider how to prevent the incorrect action or how to minimize their ability to cause an adverse event if the incorrect action was completed. A starting point for drugs administration is to explore the procedure for getting the drugs to patients and then to examine the task steps, the equipment used and the relationships between these factors. A part of the hierarchical task analysis (HTA) for the drug administration process is shown in Fig. 6.44.

HTA chart shows that the top-level goal of the system is to deliver drugs to the patient. The task steps necessary to do this are listed as tasks 1–3 on the next level

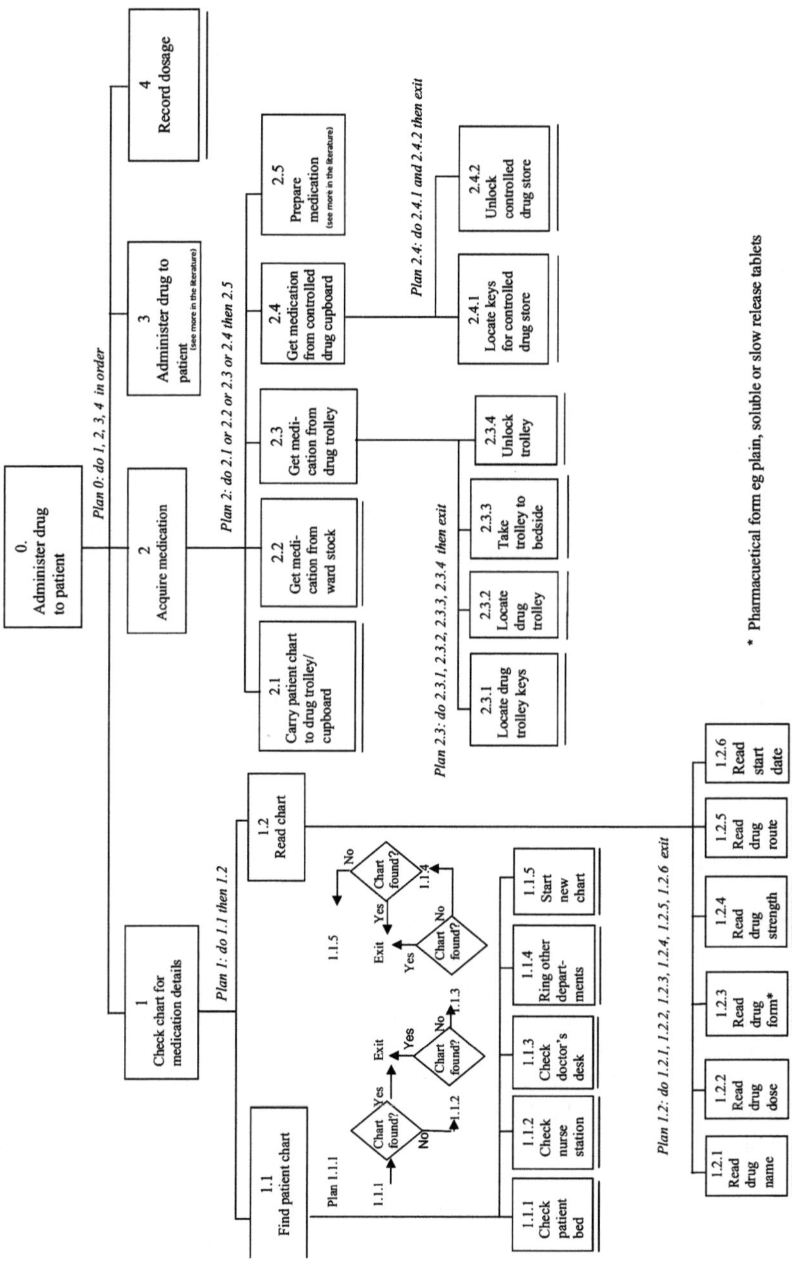

Fig. 6.44 Part of the drug administration hierarchical task analysis (Lane et al. 2006)

Table 6.31 SHERPA taxonomy of credible errors (Lane et al. 2006)

Action errors	Checking errors	Retrieval errors	Communication errors	Selection errors
A1. Operation too long/short	C1. Check omitted	R1. Information not obtained	I1. Information not communicated	S1. Selection omitted
A2. Operation mistimed	C2. Check incomplete	R2. Wrong information obtained	I2. Wrong information communicated	S2. Wrong selection made
A3. Operation in wrong direction	C3. Right check on wrong object	R3. Information retrieval incomplete	I3. Information communication incomplete	
A4. Operation too little/ much	C4. Wrong check on right object			
A5. Misalign	C5. Check mistimed			

Table 6.32 SHERPA output: human error analysis table (Lane et al. 2006)

Task step	Error mode	Description	Consequence	Recovery	P	C	Remedial measures
1.1.1	C1	Fail to check patient bed area	Chart not found: drug doses missed	1.1.2	L	M	Tagging system for location of charts

of the hierarchy. Plan 0 indicates the activities or sub-goals that should be carried out in order to achieve the goal. These activities are further broken down into operations at the lower levels. The order in which these are carried out is determined by the plan.

SHERPA uses the bottom level actions of the HTA as its inputs (task classification). These are the operations or task steps carried out to achieve the higher-level goal. The operations are evaluated for potential error using the human error taxonomy, as shown in Table 6.31.

The types of error that may occur fall into one of five behavioral categories: action, checking, retrieval, communication, and selection. Each error type in the taxonomy is coded and associated with an error mode. The task steps from the HTA are examined in turn and classified into one of the error types. The most likely error modes associated with that operation are considered. For example, the task step 1.1.1 in the HTA "check patient bed" is classified as a checking activity. Looking at the associated checking error modes in Table 6.31, only the most credible errors for the task step are taken into account.

The results of the SHERPA analysis are recorded as shown in Table 6.32. In the first column the number of the task step is listed (1.1.1). The error mode C1 is entered in the second column. This denotes a check has been missed (Table 6.32).

The third column contains an outline of the error. In this case the description would be "fail to check patient bed area". At this stage it is possible to make a prediction of what the consequence of that error might be. The chart would remain mislaid and because the nurse had no record of what drugs were due to be taken or when, drug doses would be missed. Thus, in the fourth column a description of the potential consequence of the activity is introduced. The fifth column indicates whether the error can be recovered or not. It may be that by completing further task steps, the error can be corrected. If this is the case, the task step at which the original error may be recovered is noted in the fifth column. If it is not possible to recover the error then the column is left blank.

The probability of the error occurring and its level of criticality are denoted in the table by *P* and *C* respectively. The probability of an error (ordinal probability analysis) is categorized as low (hardly ever occurs), medium (has occurred once or twice) or high (occurs frequently), while criticality (*C*) is usually either all or none and it must be acknowledged that many drug administration errors are potentially critical. However, the extent to which many administration errors cause a fatality or serious injury is highly variable and is dependent on numerous factors such as the drug's potency and therapeutic range, the age, and the condition of the patient.

For the purposes of this scenario criticality was modified to reflect three levels of severity: low (L), medium (M) and high (H). Some of the levels of severity correspond to the following descriptions:

(L) Level 0: No medication error
(M) Level 2: Error: increased need for monitoring, no change in vital sings
(H) Level 5: Error: increased monitoring and treatment, change in patient

Incidents described as low in criticality can be recoverable by an alternative course of actions. Since this version does not refer to a specific drug, it is difficult to quantify criticality. However, it is possible to derive more accurate results and quantify criticality in cases where specific drugs are used.

The last column shows the countermeasures that could be taken to reduce errors. These are mainly in the form of the design of products and technological systems (remedy analysis). It is important to note that in order to be effectively implemented, any design solution needs to regulated by appropriate management and organizational controls.

Literature indicates (Lyons et al. 2004) that apart from SHERPA, techniques such as THERP, HEART, or SLIM can also be applied to the healthcare industry. At this point the authors would like to point out that SHERPA could be also applied to the other CI domains in order to get an overall view of the role of the human failure in the corresponding domain.

6.7.7 Information and Communication Technologies

Over the past decade, governments, industry, and other interested organizations have focused on identifying, defining and coping with the possible impact of direct and indirect threats against the information and communication infrastructure (ICT), and to identify countermeasures and recommendations in addressing these vulnerabilities (Macwan 2004). Human errors happen quite often in the ICT industry, and the Network Reliability Steering Committee (USA) has published a study that shows over 40% and in some cases over 50% of network outages are attributed to human errors (Macwan 2004).

Human vulnerabilities can be divided into two main categories:

- Vulnerabilities associated with human characteristics
- Vulnerabilities in the user environment

The latter is further divided into other sub-groups such as:

- Physical; cognitive, and ethical for those which are associated with human characteristics
- Human–machine interfaces; job function-training and corporate culture for those which are related to user environment

Similar to other industries, the methodology to deal with the issue of human reliability to ICT domain includes the following task:

- Assemble a team of experts from security, risk assessment, human resources and other organizations to generate a list of vulnerabilities (based on the two main categories of human vulnerabilities) applicable to each of the specific areas of focus.
- Identify the vulnerabilities that apply to the scenarios included in risk assessment studies.
- Use one or more of a number of techniques, estimate the probability of failure associated with the vulnerabilities.
- Incorporate these results into the overall risk assessment.

Literature (Macwan 2004) indicates that techniques, such as THERP, SLIM/MAUD can be applied to the ICT industry. Although these techniques have already been described, it should be noted that to apply them to ICT scenarios requires their modification.

Risk estimation is not the only result of the analysis. In addition to that, results include recommendations to overcome human vulnerabilities such as; the better design of systems and user interface; better documentation; better training, and better staffing rotation. However malicious attacks, which is one of the most common and significant threats to the ICT industry cannot be addressed by existing HRA techniques.

6.7.8 Conclusions

CIs are particularly vulnerable given they are complex systems. The effects of malfunctions, outages or errors during their operation can lead to undesirable and catastrophic results, including; loss of communications; energy or transportation disruptions and the collapse of public health or financial services.

By developing an understanding of the causes, modes, and probabilities of human errors, valuable insights can be provided into the important characteristics of CI design. Consequently, special attention should be paid to those scenarios and to the human actions that are identified by HRA analyses, as they are crucial to the safety of CIs. Human errors are a significant threat for CIs. During the last decade HRA techniques have been applied for assessing, identifying and determining their influence on these systems. This sub-chapter described some of the most well known HRA techniques and their strengths and weaknesses are elaborated (see summarizing Table 6.18). Further developments of HRA techniques, especially concerning their use for improving the safety of CIs, include the following:

(1) The HRA techniques have been rarely used in CI scenarios and have yet to be validated. Therefore, it will take time and resources for their effective and robust use to ensure the accuracy of their results.
(2) Most of the techniques do not take into account repetitive human errors (THERP for maintenance procedures has taken those errors into account). They consider that an error can happen only once per task or scenario.
(3) These techniques rely to a large extent on expert judgment. Consequently, the results obtained may differ, depending upon the analysts' background and experience.
(4) The techniques are task oriented. Hence, cases or tasks with no formal protocol require time-consuming resources in order to achieve a high level detail, and are also not easily applied by analysts. In addition, good practices include the observation of actual operation and recommend not relying only on the procedural description of the tasks.
(5) There is lack of data on human errors or incidents that occur in complex systems due to human errors. In order to overcome this problem, data from simulators are used. However, it is difficult to calibrate simulator data and correlate them with real-world performance.
(6) Not all of HRA techniques take into consideration the different impact of PSFs onto human performance. When they do, they depend on experts' judgment. In addition the interdependencies between PSFs are also not addressed in most of the techniques.
(7) Even in the most recent HRA techniques, a number of complex PSFs, such as the organisational factors, the quality of the procedures, or cultural differences are not adequately taken into account.

Finally, it should be noted that terrorist attacks threaten the safe function of CIs. Although, terrorist attacks are caused by humans, they are not considered as human errors rather as violations. Therefore, they are not yet examined by HRA techniques.

References

Albert R, Barabási A-L (2002) Statistical mechanics of complex networks. Rev Mod Phys 74:47–97

Albert R, Jeong H, Barabási A-L (2000) Error and attack tolerance of complex networks. Nature 406:378–382

Apostolakis GE, Lemon DM (2005) A screening methodology for the identification and ranking of infrastructure vulnerabilities due to terrorism. Risk Anal 25:361–376

Argonne National Laboratory (2008) The electricity market complex adaptive system (EMCAS), technical report. Argonne, USA

Batagelj V (1994) Semirings for social networks analysis. J Math Sociol 19(1):53–68

Beeker ER, Page EH (2006) A case study of the development and use of a MANA-based federation for studying US border operations. In: Proceedings of the 38th conference on winter simulation, 3–6 Dec. 2006, pp 841–847

Behdani B, Lukszo Z et al (2010) Performance analysis of a multi-plant specialty chemical manufacturing enterprise using an agent-based model. Comput Chem Eng 34(5):793–801

Bell J, Holroyd J (2009) Review of human reliability assessment methods. Health & Safety Laboratory

Billinton R, Li W (1994) Reliability assessment of electrical power systems using Monte Carlo methods. Plenum Press, New York

Birolini A (2007) Reliability engineering: theory and practice. Springer, Berlin

Boccaletti S, Latora V, Moreno Y, Chavez M, Hwang D-U (2006) Complex networks: structure and dynamics. Phys Rep 424:175–308

Bonabeau E (2002) Agent-based modeling: methods and techniques for simulation human systems. Proc Natl Acad Sci USA 99:7280–7287

Buchanan M (2009) Meltdown modeling. Nature 460:680–682

Cadini F, Zio E, Petrescu C-A (2009) Using centrality measures to rank the importance of the components of a complex network infrastructure. In: Critical information infrastructure security, proceedings of the 3rd international workshop on critical information infrastructures security, CRITIS 2008, Rome, Italy, 13–15 October 2008, pp 155–167

Cardellini V, Casalicchio E, Galli E (2007) Agent-based modelling of interdependencies in critical infrastructures through UML. In: Proceedings of the 2007 spring simulation multiconference, Norfolk, Virginia, USA

Chen J, Thorp S-J, Dobson I (2005) Cascading dynamics and mitigation assessment in power system disturbances via a hidden failure model. Int J Electr Power Energ Syst 27:318–326

Chien SH, Dykes AA, Stetkar JW, Bley DC (1988) Quantification of human error rates using a SLIM-based approach. In: IEEE fourth conference on human factors and power plants, June 5–9, 1988, Monterey, California

Coffman EG, Ge Z, Misra V, Towsley D (2002) Network resilience: exploring cascading failures within BGP. In: Proceedings of the 40th annual Allerton conference on communications, computing and control

Crucitti P, Latora V, Marchiori M, Rapisarda A (2003) Efficiency of scale-free networks: error and attack tolerance. Physica A 320:622–642

D'Inverno M, Luck M (2004) Understanding agent systems. Springer, Berlin

Dahmann JS, Fujjimoto RM, Weatherly RM (1997) The department of defense high level architecture. In: Proceedings of the 29th conference on winter simulation, Atlanta, Georgia, United States: IEEE Computer Society

Darby J (2006) Evaluation of terrorist risk using belief and plausibility. PSAM8, New Orleans, USA

Department of Energy (2005) From OE-417, electric emergency incident and disturbance report. US Department of Energy, Office of Electricity Delivery and Energy Reliability. http://www.eia.doe.gove/oss/forms.html. Accessed date/month/year

Dobson I, Carreras AB, Lynch VE, Newman DE (2004) Complex systems analysis of series of blackouts: cascading failure, criticality, and self-organization. In: Bulk power system dynamics and control—VI. Cortina d'Ampezzo, 22–27 August 2004

Dobson I, Carreras BA, Newman DE (2005) A loading-dependent model of probabilistic cascading failure. Probab Eng Inform Sci 19(15):32

Dobson I, Carreras BA, et al (2007) Complex systems analysis of series of blackouts: cascading failure, critical points, and self-organization. Chaos 17(2): 026103 (13 pages), 2007

Duflos S, Diallo A, Grand AGL (2007) An overlay simulator for interdependent critical information infrastructures. In: Proceedings of the 2nd international conference on dependability of computer systems, IEEE computer society

Embrey DE, Kirwan B (1983) A comparative evaluation of three subjective human reliability quantification techniques. In: Proceedings of the annual ergonomics society conference

Embrey DE, Humphreys PC, Rosa EA, Kirwan B, Rea K (1984) SLIM-MAUD: An approach to assessing human error probabilities using structured expert judgement. United States Nuclear Regulatory Commission

Erdős P, Rényi A (1960) On the evolution of random graphs. Publ Math Inst Hung Acad Sci A 5:17–61

Eurocontrol (2007) Overview of HRA methods. Farandole project

Eusgeld I, Dietz S (2009) "System-of-systems" approach for interdependent critical infrastructures. In: Proceedings of the annual conference of the European safety and reliability association (ESRA), Prague

Eusgeld I, Nan C (2009) Creating a simulation environment for critical infrastructure interdependencies study. In: Industrial engineering and engineering management, 2009. IEEM 2009. IEEE international conference on, 8–11 December 2009, pp 2104–2108

Eusgeld I, Kröger W, Sansavini G, Schläpfer M, Zio E (2009) The role of network theory and object-oriented modeling within a framework for the vulnerability analysis of critical infrastructures. Reliab Eng Syst Safe 94(5):954–963

Ezel BC (2007) Infrastructure vulnerability assessment model (I-VAM). Risk Anal 27:571–583

Floyd RW (1962) Algorithm 97: shortest path. Commun ACM 5(6):345

Forester J, Ramey-Smith A, Bley D, Kolaczkowski A, Cooper S (1998) Discussion of comments from a peer review of a technique for human event analysis (ATHEANA)

Forester J, Kolaczkowski A, Lois E, Kelly D (2006) Evaluation of human reliability analysis methods against good practices. United States Nuclear Regulatory Commission

Forester J, Kolaczkowski A, Cooper S, Bley D, Lois E (2007) ATHEANA user's guide. United States Nuclear Regulatory Commission

Freeman LC (1979) Centrality in social networks conceptual clarification. Soc Networks 1(3):215–239

Garrick BJ, Hall JE, Kilger M, McDonald JC, McGroddy JC, O'Toole T, Probst PS, Rindskopf Parker E, Rosenthal R, Trivelpiece AW, Van Arsdale L, Zebroski E (2004) Confronting the risks of terrorism: making the right decisions. Reliab Eng Syst Safe 86:129–176

Garrido JM (2009) Object-oriented simulation. Springer, Berlin

Gibson HW, Kirwan B (2008a) Application of the CARA HRA tool to air traffic management safety cases. In: Proceedings of the 9th international conference on probabilistic safety assessment and management (PSAM9)

Gibson HW, Kirwan B (2008b) Current trends in human reliability assessment. In: Proceedings of the international conference on contemporary ergonomics (CE2008), Nottingham, UK, April 1–3, 2008

Gorbil G, Gelenbe E (2009) Design of a mobile agent-based adaptive communication middleware for federations of critical infrastructure simulations. In: Proceedings of CRITIS 2009

Grigg NS (2003) Water utility security: multiple hazards and multiple barriers. J Infrastruct Syst 9(2):81–88

Grigg C, Wong P et al (1996) The IEEE reliability test system 1996, 1996 IEEE/PES winter meeting, Baltimore, Maryland, IEEE-Inst Electrical Electronics Engineers Inc

Grimm V, Revilla E et al (2005) Pattern-oriented modeling of agent-based complex systems: lessons from ecology. Science 310:987–991

Haimes YY, Horowitz BM (2004) Modeling interdependent infrastructures for sustainable counterterrorism. J Infrastruct Syst 10:33–42

Hansen JV, Lowry PB, Meservy RD, McDonald DM (2007) Genetic programming for prevention of cyber terrorism through dynamic and evolving intrusion detection. Decis Supp Syst 43(4):1362–1374

Hines P, Blumsack S (2008) A centrality measure for electrical networks. In: Proceedings of the 41st Hawaii international conference on system science

Hines P, Apt J, Talukdar S (2008) Trends in the history of large blackouts in the United States. IEEE, Authorized licensed use limited to ETH Bibliothek Zurich

Hollangel E (1998) Cognitive reliability and error analysis method. Elsevier, New York

Hopkinson KM, Giovanini R, Wang XR (2003) Integrated commercial off-the-shelf software for agent-based electric power and communication simulation. In: Proceedings of the 2003 winter simulation conference, pp 1158–1166

Huang K (1987) Statistical mechanics, 2nd edn. John Wiley & Sons, New York, pp 31–35 206–210

Hudson LD, Ware BS, Laskey KB, Mahoney SM (2002) An application of Bayesian networks to antiterrorism risk management for military planners, Rapport technique. George Mason University

IEEE (1999) IEEE RTS Task Force of APM Subcommittee. The IEEE reliability test system 1996. IEEE Trans Power Syst 14, 1010–1020

IEEE (2000) IEEE standard for modeling and simulation, high level architecture (HLA): framework and rules, IEEE Std. 1516–2000, i–22

IEEE (2009) IEEE draft standard for modeling and simulation (M&S) high level architecture (HLA): framework and rules, IEEE unapproved draft std P1516/D5

IRGC (2006) Managing and reducing social vulnerabilities from coupled critical infrastructures, White Paper No. 3, International Risk Governance Council, Geneva, p. 68

Jennings RN (2000) On agent-based software engineering. Artif Intell 117:277–296

Kaegi M, Mock R, Kröger W (2009) Analyzing maintenance strategies by agent-based simulations: a feasibility study. Reliab Eng Syst Safe 94:1416–1421

Kaplan S, Garrick BJ (1981) On the quantitative definition of risk. Risk Anal 1:11–27

Kim IK, Ma YB, Lee JS (2006a) Adaptive quantization-based communication data management for high-performance geo-computation in grid computing. In: Proceedings of the grid and cooperative computing workshops, 2006. GCCW '06, fifth international conference on, Oct 2006, pp 470–476

Kim J, Jung W, Jang S, Wang J (2006b) A case study for the selection of a railway human reliability analysis method. In: International railway safety conference, 22–27 October 2006, Belfast

Kirwan B (1994) A guide to practical human reliability assessment. Taylor & Francis, London

Kirwan B, Gibson HW (2007) CARA: a human reliability assessment tool for air traffic safety management: technical basis and preliminary architecture. In: Proceedings of the fifteenth safety-critical systems symposium

Koonce AM, Apostolakis GE, Cook BK (2008) Bulk power risk analysis: ranking infrastructure elements according to their risk significance. Int J Electr Power Energ Syst 30:169–183

Kröger W (2005) Risk analyses and protection strategies for operation of nuclear power plants, in Landolt-Börnstein New Series Vol. VIII/3B (advanced materials and technologies/energy). Springer-Verlag, Berlin

Kröger W (2008) Critical infrastructures at risk: a need for a new conceptual approach and extended analytical tools. Reliab Eng Syst Safe 93:1781–1787

Kyriakidis MA (2009a) A scoping method for human performance integrity and reliability assessment in process industries. Laboratory for Safety Analysis, Institute for Energy Technology, D-MAVT, ETH Zurich

Kyriakidis MA (2009b) A study regarding human reliability within power system control rooms. Laboratory for Safety Analysis, D-MAVT, ETH Zurich

Lane R, Stanton NA, Harrison D (2006) Applying hierarchical task analysis to medication administration errors. Appl Ergon 37:669–679

Latora V, Marchiori M (2001) Efficient behavior of small-world networks. Phys Rev Lett 87(19):198701 (1–4)

Latora V, Marchiori M (2005) Vulnerability and protection of infrastructure networks. Phys Rev E 71:015103 (1–4)

Latora V, Marchiori M (2007) A measure of centrality based on the network efficiency. New J Phys 9:188

Lees M, Logan B, Theodoropoulos G (2007) Distributed simulation of agent-based systems with HLA. ACM Trans Model Comput Simul 17(3):11

Lemmers AJJ, Kuiper PJ, Verhage FR (2002) Performance of a component-based flight simulator architecture using the HLA paradigm. In: Proceedings of the AIAA modeling and simulation technologies conference and exhibit. California, USA

Lempert R (2004) Robust decision making. HDGC Seminar, February 2004

Lieshout FV, Cornelissen F, Neuteboom J (2008) Simulating rail traffic safety systems using HLA 1516, Atos origin technical automation

Lyons M, Adams S, Woloshynowych M (2004) Ch. Vincent, human reliability analysis in healthcare: a review of techniques. Int J Risk Saf Med 16:223–237

Macal CM, North MJ (2005) Tutorial on agent-based modeling and simulations. In: Proceedings of the 2005 winter simulation conference. Orlando FL, USA

Macwan A (2004) Approach for identification and analysis of human vulnerabilities in protecting telecommunications infrastructure. Bell Labs Tech J 2:85–89

Michaud D, Apostolakis GE (2006) Screening vulnerabilities in water supply networks. J Infrastruct Syst 12:230–242

Möller B, Löfstrand B, Lindqvist J, Backlund A, Waller B, Virding R (2005) Gaming and HLA 1516 interoperability within the Swedish defense. In: Proceedings of the 2005 fall simulation interoperability workshop

Möller B, Morse KL, Lightner M, Little R, Lutz R (2008) HLA evolved: a summary of major technical improvements

Moore AD (2006) Application of the API/NPRA SVA methodology to transportation security issues. J Hazard Mater 130:107–121

Morse KL, Lightner M, Little R, Lutz B, Scrudder R (2006) Enabling simulation interoperability. Computer 39:115–117

Motter AE (2004) Cascade control and defense in complex networks. Phys Rev Lett 93(9):098701 (1–4)

Motter AE, Lai YC (2002) Cascade-based attacks on complex networks. Phys Rev E 66:065102 (1–4)

Nan C, Eusgeld I (2011) Adopting HLA standard for interdependency study. Reliab Eng Syst Safe 96(1):149–159

Newman MEJ, Girvan M (2004) Finding and evaluating community structure in networks. Phys Rev E 69(2):026113

Newman DE, Nkei B, Carreras BA, Dobson I, Lynch VE, Gradney P (2005) Risk assessment in complex interacting infrastructure systems. In: Proceedings of the 38th Hawaii international conference on system sciences

Nieminen J (1974) On the centrality in a graph. Scand J Psychol 15(1):332–336

Paté-Cornell ME, Guikema S (2002) Probabilistic modeling of terrorist threats: a systems analysis approach to setting priorities among countermeasures. Mil Oper Res 7:5–20

Patterson SA, Apostolakis GE (2007) Identification of critical locations across multiple infrastructures for terrorist actions. Reliab Eng Syst Safe 92:1183–1203

Pederson P, Dudenhoeffer D, Hartly S, Permann M (2006) Critical infrastructure interdependency modeling: a survey of US and international research. Idaho National Laboratory

Piwowar J, Chatelet E, Laclemence P (2009) An efficient process to reduce infrastructure vulnerabilities facing malevolence. Reliab Eng Syst Safe 94:1869–1877

Rehtanz C (2003) Autonomous systems and intelligent agents in power system control and operation. Springer, Berlin

Robles RJ, Choi M-K, Coh E-S, Kim S-S, Park G-C, Lee J-H (2008) Common threats and vulnerabilities of critical infrastructures. International Journal of Control and Automation (1) 17–22.

Rosato V, Bologna S, Tiriticco F (2007) Topological properties of high-voltage electrical transmission networks. Electr Pow Syst Res 77:99–105

Rosato V et al (2008) A complex system's view of critical infrastructures. In: Helbing D (ed) Understanding complex systems. Springer, Berlin, Heidelberg, pp 241–260

Ross TJ (2004) Fuzzy logic with engineering applications, 2nd edn. Wiley, New York

Sabidussi G (1966) The centrality index of graphs. Psychometrika 31(4):581–603

Schläpfer M, Kessler T, Kröger W (2008) Reliability analysis of electric power systems using an object-oriented hybrid modeling approach. In: Proceedings of the 16th power systems computation conference, Glasgow

Stanton N, Hedge A, Brookhuis K, Salas E, Hendrick H (eds) (2005) Handbook of human factors and ergonomics methods. CPC Press, New York

Strogatz SH (2001) Exploring complex networks. Nature 410:268–276

Swain AD, Guttmann HE (1983) Handbook of human reliability analysis with emphasis on nuclear power plant applications. United States Nuclear Regulatory Commission

TERNA (2002) Dati statistici sull'energia elettrica in Italia. Technical report. Terna S.p.A.—Rete Elettrica Nazionale (in Italian) http://www.terna.it/LinkClick.aspx?fileticket=PUvAU57MlBY%3d&tabid=418&mid=2501. Accessed 08/06/2011

US Department of Defense (1998) High level architecture interface specification. DOD

US Department of Defense (2000) High level architecture run-time interface programmers guide. DOD

US Department of Defense (2007) DOD Directive 4603.05: interoperability and supportability of information technology (IT) and national security systems. DOD

Vespignani A (2009) Predicting the behaviour of techno-social systems. Science 325:425–428

VSE-AES Statistik (2005) Statistik 2500 über die Verfügbarkeit der Elektrizitätsversorgung der Schweiz

Wasserman S, Faust K (1994) Social networks analysis. Cambridge University Press, Cambridge

Watts DJ (1999) Networks, dynamics, and the small-world phenomenon. Am J Sociol 105(2):493–527

Watts DJ (2002) A simple model of global cascades on random networks. PNAS 99(9):5766–5771

Watts DJ, Strogatz SH (1998) Collective dynamics of 'small-world' networks. Nature 393:440–442

Wood AJ, Wollenberg BF (1996) Power generation, operation and control. John Wiley & Sons, New York

Wreathall J, Nemeth C (2004) Assessing risk: the role of probabilistic assessment (PRA) in patient safety improvement. Qual Saf Health Care 13:206–212

Wreathall J, Roth E, Bley D, Multer J (2003) Human reliability analysis in support of risk assessment for positive train control. United States Department of Transportation

Yilmaz L, Ören T (2009) Agent-directed simulation and systems engineering. Wiley, New York

Zacharewicz G, Alix T, Vallespir B (2009) Services modeling and distributed simulation DEVS/HLA supported. In: Proceedings of the 2009 winter simulation conference (WSC), 13–16 December 2009, pp 3023–3035

Zhao Z, Albada DV, Sloot P (2005) Agent-based flow control for HLA components. Simulation 81:487–501

Zimmermann R (2001) Social implications of infrastructure network interactions. J Urban Technol 8(3):97–119

Zimmermann R (2004) Decision-making and the vulnerability of interdependent critical infrastructure. In: Proceedings of the IEEE international conference on systems, man, and cybernetics, the Hague, Netherlands

Zio E (2007a) From complexity science to reliability efficiency: a new way of looking at complex network systems and critical infrastructures. Int J Critical Infrastructures 3:488–508

Zio E (2007b) An introduction to the basics of reliability and risk analysis, vol 13., Series on quality, reliability and engineering statisticsWorld Scientific, Singapore

Zio E, Golea LR (2010) Analyzing the topological, electrical and reliability characteristics of a power transmission system for identifying its critical elements. Reliab Eng Syst Safe(submitted)

Zio E, Sansavini G (2007) A systematic procedure for analyzing network systems. Int J Critical Infrastructures 4(1–2):172–184

Zio E, Sansavini G (2008) Modeling failure cascades in networks systems due to distributed random disturbances and targeted intentional attacks. In: Martorell et al. (eds) Safety, reliability and risk analysis: theory, methods and applications. Proceedings of ESREL 2008 and 17th SRA Europe annual conference, Valencia, Spain, Taylor & Francis Group, London, 22–25 September 2008

Zio E, Sansavini G (2011a) Component criticality in failure cascade processes of network systems. Risk Anal. doi:10.1111/j.1539-6924.2011.01584.x

Zio E, Sansavini G (2011b) Modeling interdependent network systems for identifying cascade-safe operating margins. IEEE Trans Reliab 60(1):94–101

Zio E, Sansavini G, Maja R, Marchionni G (2008) An analytical approach to the safety of road networks. Int J Reliab Qual Saf Eng 15(1):67–76

Kardes E (2005) Robust Stochastic Games and Applications to Counter-Terrorism Strategies. Center for Risk and Economic Analysis of Terrorism Events, University of Southern California, Los Angeles CA.

Subotic B (2007) Framework for the analysis of controller recovery from equipment failures in air traffic control, Civil and Environmental Engineering, Imperial College, London

Chapter 7
Concluding Considerations

In this book, the subject of vulnerable systems has been addressed in depth. The focus has been on engineered physical critical infrastructures (CIs) such as the networks for energy supply, transportation, information and telecommunication. These are large-scale arrays of systems and assets that function systematically to produce and distribute services vital for modern economy and social welfare.

One main motif underlying the viewpoints presented in the book is that of the complexity of these systems and the related emergent behaviors which may arise in collective ways, difficult to predict from the superposition of the behavior of the individual elements of the system. This complex, emergent behavior has been demonstrated by system breakdowns often triggered by small perturbations followed by accelerating cascades and large-scale, border-crossing consequences, stressing the importance of (inter)dependencies (see Table 2.4 for details). As a conclusion, the analysis of these systems cannot be carried out with classical analytical methods of system decomposition and logic analysis; a framework of analysis is needed to integrate a number of methods capable of viewing the problem from different perspectives (topological and functional, static and dynamic, etc.).

Another relevant complexity attribute that has been stressed throughout relates to the increased integration of CIs, e.g., driven by the pervasive use of computer-based communication and control systems, which is beneficial in many respects but on the other hand it introduces additional vulnerabilities due to the increased (inter)dependence which can lead to surprising behaviors in response to perturbations. The electric power supply system, for instance, is undergoing a design re-conceptualization to allow for the integration of large shares of electricity produced by harvesting solar and wind energies at the most suitable sites (e.g., desert solar and offshore wind farms). The grids will become "smarter" by decentralized generation, smart metering, new devices for increased controllability, etc., which will "convert the existing power grid from a static infrastructure to be operated as designed into a flexible, adaptive infrastructure operated proactively" (IEC 10). These types of developments can be expected to be shared

W. Kröger and E. Zio, *Vulnerable Systems*, DOI: 10.1007/978-0-85729-655-9_7,

among CIs, but the final states and related challenges may be different. For instance, the extended use of traffic control systems may increase the complexity of our road transport systems at a first glance, but may in the end turn them into less dynamic systems upon transferring important influencing factors such as speed, distance keeping, performance, etc., from individual drivers to computer-based systems.

Another important aspect that has underpinned a number of considerations in this book, relates to the fact that the types of CIs considered are subject to a broader spectrum of hazards and threats, including malicious attacks; this has led to the conclusion that an all-hazards approach is required for the understanding of the failure behavior of such systems, for their consequent protection.

The above considerations have led to the structuring of the analysis of vulnerable CI systems in the book into five levels:

- The *problem definition level*, at which the concepts of CI and vulnerability have been unambiguously introduced, together with and in reference to other key terms.
- The *issues level*, at which the peculiar characteristic features of such intra- and inter-connected systems have been described, the related challenges with respect to their analysis have been identified and the methods have been critically addressed.
- The *approaches level*, at which approaches to meeting the challenges have been discussed.
- The *methods level*, at which the methods currently under consideration for the analysis of vulnerable CIs have been illustrated to a level of detail sufficient for their implementation and use, and with a careful analysis of their respective strengths, limitations, and degrees of maturity for practical application.
- The *framework level*, at which an integrated view of the procedure for vulnerability analysis has been offered, coherent with the relevant aspects and features of intra- and inter-dependent CI systems highlighted above.

As unidirectional and bidirectional relationships within and among CIs are expected to continue standing as a real-world issue and representing a major challenge for analysis methods, the evaluation of the vulnerability of CIs becomes central for their future efficient and safe development and operation. As a contribution to this, the present book introduces an operational definition of vulnerability in terms of weaknesses that render the CI susceptible to destruction or incapacitation when exposed to a set of hazards or threats (see Fig. 1.3).

Furthermore, in practice CIs are considered critical with regard to some criteria, possibly varying from country to country and from one perspective to another, e.g., corporate, societal, system or user's view. In the book, an attempt has been made to distinguish the degrees of criticality, because this is considered to help guiding the vulnerability analysis.

Turning the attention to vulnerability analysis, the book embraces the classical characterization of CIs as systems made of a large number of elements interacting in different complex ways which make system behaviors emerge (sometimes in unexpected ways). The differentiation from complicated systems is clearly pointed out (see also Table 2.1), together with the resulting fact that decomposition may not help in identifying and analyzing the (emergent) system behaviors, so that the CI system itself must be looked at as a whole with all its intra- and inter-dependencies.

In the book, the procedure of vulnerability analysis offered comprises three tasks (on top of the analysis of the system structure, properties and (all) hazards): the quantification of vulnerability indicators, the identification of critical elements and the application of the analysis results for system improvements (see Fig. 3.1). Correspondingly, the challenges to the methods used within the vulnerability analysis procedure depend on the specific objectives of the analysis and on the system characteristics; a common challenge comes from the large number of parameters needed to characterize the model of the system and the paucity of reliable data in support. Other specific challenges come from the need to capture the emergent behaviors and intricate rules of interaction, various system features like multi-layering, state changes, adaptation to new developments, "system-of-systems" characteristics, the susceptibility to a broad spectrum of hazards and threats (see Fig. 3.3). All these features need to be tackled by the methods for vulnerability assessment. A number of these have been compiled, characterized, and critically evaluated against the assessment steps, the desired outputs and practical objectives (Table 4.1). Methods of statistical analysis, probabilistic modeling techniques (e.g., Bayesian networks), and tabular methods for hazard studies and risk analysis (HAZOP, FMEA, pure expert judgment) used in isolation have been concluded to show limited chances of success against the challenges posed; on the contrary, integration of elements of probabilistic risk assessment (e.g., adapted logic trees), complex network theory and agent-based modeling and simulation techniques, including high-level architecture (HLA), appear most promising.

In the end, it is concluded that a universal, all encompassing modeling approach capable of addressing by itself the assessment of vulnerable systems does not exist. Rather, a conceptual framework is proposed (Fig. 5.1) to systematically tackle the problem in a step-wise and integrated fashion which stands on a preparatory phase, a screening analysis and an in-depth analysis of major critical areas of the system. Integration is intended to refer also to the multitude of perspectives of the different "players" (users/operators) involved, their different logics, and even potential confidentiality issues.

The book points at methods of network theory, probabilistic risk analysis, cascading failure dynamics for the screening analysis, which although performed at a relatively high level of abstraction is capable of providing generic insights into the topology and phenomena involved which can serve as guidance for the successive in-depth vulnerability analysis phase. For this latter more detailed phase, the combination of agent-based modeling with Monte Carlo simulation techniques and, where appropriate, with physical models (e.g., for mass or power flows) has been pointed at in the book as the most promising approach.

For (inter)dependency analysis HLA standard might be attractive as the modeling of the interconnected systems in a super-model with adequate granularity might be impossible due to further exacerbated computer time/resources issues and overall model complexities.

In the book, the above mentioned methods of screening and in-depth analysis have been illustrated by a detailed description of their functioning and evaluated as to how/to what extent they succeed in the analysis of vulnerable systems and in meeting the related challenges, and to how mature they are for practical application. The considerations that have merged in the book (see Table 7.1) show in general terms differing levels of capabilities; some of the methods have been developed in other fields of technical risk assessment, e.g., PRA for nuclear power plants, or other sectors, e.g., network theory and agent-based modeling in social systems, and applied/adapted to the vulnerability assessment of engineered CIs. The results gained and the way to illustrate them also differ (e.g., metrics of robustness of network topology, curves depicting frequency, and size of potential blackouts) and are, therefore, differently qualified as input for decision making. Although some of the concepts and results have been proved against statistical data (e.g., statistical data on blackouts) and benchmarked against other methods, the analysis provided in the book shows that there is still general lack of validation, there are still contradictions between (limited) empirical investigations and theoretical analysis.

It is worth mentioning that the specific advantages and limitations of a framework of analysis laying over two levels of system abstraction (high-level for screening and detailed for in-depth analysis) eventually lay the groundwork for their specific and separate application to particular areas as well as for their interplay within a comprehensive analysis of CI networks. Highly detailed modeling approaches with close adherence to reality may also serve for quantifying the reliability of a specific system under given operational conditions. In complementation, minimal, highly abstract models allow for a qualitative identification of basic underlying mechanisms and generic key factors, which are not restricted to specific systems. Such insights may again serve as valuable clues on where to put the focus with highly detailed modeling approaches. On the other way around, quantitative results from specific systems may also be checked with regard to their generality by using minimal models.

Finally, uncertainty analysis is much needed for inclusion into the picture, although still rarely an element of an overall vulnerability assessment; in this respect, computational tools for uncertainty propagation and sensitivity analysis in large scale applications, with reasonable computing times are missing. The same holds for the need of including the interactions and influences of the human operators, a cross-cutting issue into vulnerability analysis which has been addressed in other than CI sectors, mainly in the framework of PRA for nuclear power plants, but is now starting to arise also in the field of vulnerable CIs analysis.

In conclusion, the writing of the book has served the authors to build the confirmation that there is still a gap between the ability to design and operate complex networked CI systems and the ability to understand, model and simulate them for identifying and reducing vulnerabilities. In this view, intensive research, development and application work must be continued and many other books can be expected in future.

Table 7.1 Summary of the various methods for the vulnerability analysis of CIs

Technique	Capabilities			Accountancy of uncertainties	Maturity for practical applications		
	Suitability	Advantage	Disadvantage		Degree of validation[a]	User friendly tools	Presentation of results
Complex network theory	Mainly for capturing topological features of the networked CI; addition of "weights" can account for physical attributes of the CI components, to a certain extent	Simple and fast analysis, which provides quantitative indicators of the topological characteristics of the network of connections underpinning a CI. Allows to identify topologically-critical areas of the system	If used in isolation, it can provide only indications on the topology of the system, potentially failing to represent the other features related to the flow-driving characteristics	In the weighted scheme of analysis, the possibility of accounting for uncertainties on the weighting parameters seems straightforward and it does not pose particular computational challenges given that the evaluation of the underlying (topological) model is very fast	A number of validation proofs have been offered with regard to the topological characterization of real networked infrastructures	Very limited efforts have been made in providing user friendly tools to carry out complex network analyses	The results are typically synthesized in distributions and point-value metrics which summarize topological features of the network connection Web

(continued)

Table 7.1 (continued)

Technique	Capabilities			Accountancy of uncertainties	Maturity for practical applications		
	Suitability	Advantage	Disadvantage		Degree of validation[a]	User friendly tools	Presentation of results
Risk analysis	Mainly for parts of a CI. Extended application to an entire CI would typically require excessive resources	Allows a systematic and logic analysis of vulnerability scenarios which may affect a CI, while at the same time allowing for their quantification of likelihood of occurrence and consequence, and for the identification of the critical CI components	The efforts in logic modeling and quantification are significant. The capability of providing an exhaustive analysis is limited, particularly in view of the unexpected emergent behaviors and of the many (inter) dependencies	The quantification part of the analysis is typically based on probabilistic modeling, which is naturally suited for uncertainty analysis. The computational burden depends on the extent of the modeling upon which uncertainty analysis is to be mounted, with its additional computational burden	Validation of such analyses is quite difficult to obtain. Quality assurance is needed in support, but this is far from trivial	A number of tools have been developed, but not specific for CI vulnerability analysis. Adaptation would be needed, whose effort and degree of success is difficult to foresee	Results are provided in terms of logic scenario trees, risk matrices and distributions. These are quite familiar representations for risk analysts

(continued)

Table 7.1 (continued)

Technique	Capabilities			Accountancy of uncertainties	Maturity for practical applications		
	Suitability	Advantage	Disadvantage		Degree of validation[a]	User friendly tools	Presentation of results
Probabilistic modeling of cascades dynamics	Allows superposing the cascade dynamic process onto the topology of the system, within a relatively high-level of abstraction. In this sense, it is suitable for complementing the topological analysis with dynamic features representative of the cascade phenomena which may occur	The high-level of abstraction leads to a modeling which is still sufficiently slim to allow running what-if scenarios at reasonable computational expenses. The results of this add to the topological analysis by confirming or not certain network connectivity characteristics and critical points, in light of also the dynamic patterns of cascade evolution	Although dynamics is added to complement the topological analysis, it may still fail to capture physical aspects of the cascade dynamics related to the flow-driven processes actually occurring	Uncertainties on the development of the cascade scenarios can be added in a quite straightforward manner, with no particular additional burden of computations	A number of validation tests have been presented with regard to the capability of modeling the main characteristics of real cascading processes, e.g., electrical blackouts	Very limited efforts have been made to provide user-friendly tools	Results are usually provided in terms of network connectivity characteristics and cascade sizes, as a function of key system parameters. From these, phase transitions are sought which testify the emergence of new behaviors and allow dimensioning protections and preventions with respect to the escalation of uncontrolled cascades

(continued)

Table 7.1 (continued)

Technique	Capabilities			Accountancy of uncertainties	Maturity for practical applications		
	Suitability	Advantage	Disadvantage		Degree of validation[a]	User friendly tools	Presentation of results
Agent-based modeling	Allows simulation of complex systems, comprising a large number of heterogeneous components interacting with each other	Close adherence to reality; capable to capture highly non-linear dynamics and emergent phenomena; integration of non-technical system elements	Usually requires a large number of model parameters and may imply long computational times	The level of modeling details allows to quantify a multitude of time-dependent reliability and vulnerability aspects, and thus potentially provides essential insights into the diverse sources of uncertainties	While it has been widely validated in fields like social sciences, it still lacks comprehensive feasibility studies in the field of CIs	A number of user friendly tools are available	Due to its high flexibility, it offers a broad range of quantitative result representations. Examples include disruption times or “frequency–consequence (F–C)” diagrams

(continued)

Table 7.1 (continued)

Technique	Capabilities			Accountancy of uncertainties	Maturity for practical applications		
	Suitability	Advantage	Disadvantage		Degree of validation[a]	User friendly tools	Presentation of results
High-level architecture	A simulation standard mainly for representing multiple interacting systems such as CIs in one simulation tool, which is not suitable to all simulation developments. A "pre-investigation" is recommended before adopting this standard	As a promising approach for integrating different modeling methods, the distribution of simulation components improves the flexibility/reusability of the simulation tool and decreases its overall complexity	It is not a "plug-and-play" standard. Resources and time required to implement an HLA-compliant simulation tool could be significant	Synchronization issues between distributed simulation components, which means that one component could process events forwarded by other components in non-increasing time stamp order, can account for uncertainties on overall simulation	Validation of such approach is difficult. Extra validation experiments need to be developed and conducted, which should include the validation experiments regarding the decomposition of simulation components	Several commercial and free ready-to-use software tools have been developed and are available to users who try to adopt HLA standard	Results are usually represen-tations of the integration of model outputs from each distributed simulation components and corre-sponding systematic analyses

(continued)

Table 7.1 (continued)

Technique	Capabilities			Accountancy of uncertainties	Maturity for practical applications		
	Suitability	Advantage	Disadvantage		Degree of validation[a]	User friendly tools	Presentation of results
Human reliability and performance modeling	Methods are used to assess human performance during the execution of a task in a qualitative matter and to estimate the corresponding human error probability	With a relatively simple and straightforward manner allows analysts to estimate human contribution to the risk, to calculate the possibility of the occurrence of a human error and to identify the factors that compromise human performance. In addition suggestions either to improve safety or to deal with the unpleasant consequences are provided	Lack of data and biased experts judgment can compromise the validity of the analysis. Findings may vary based on analysts experience, background and training. Complex techniques can only be used by experienced and trained analysts. The analysis can be both time-consuming and exhaustive	Uncertainty analysis was not included in the first stages of human performance modeling. Both interactions between employees and interdependencies between factors that compromise human performance were not taken into account, Only recently, new approaches are trying to incorporate uncertainty analysis into the overall human performance modeling frame	A number of HRA techniques have been applied to several domains, such as nuclear power plants, chemical plants, medicine and transportation. However, the lack of data may compromise the validity of the methods	Despite the large number of existing HRA techniques only few tools have been developed to facilitate their use. Nevertheless, most of the methods can be implemented by using pen and paper	Results are usually a combination of quantitative and qualitative outcomes. In the first case the human error probability is estimated. Qualitative results provide to analysts a detailed illustration of the possible errors

[a] Including proof of concepts, benchmarking against other methods, validation

MIX
Papier aus verantwortungsvollen Quellen
Paper from responsible sources
FSC® C105338

If you have any concerns about our products,
you can contact us on
ProductSafety@springernature.com

In case Publisher is established outside the EU,
the EU authorized representative is:
Springer Nature Customer Service Center GmbH
Europaplatz 3, 69115 Heidelberg, Germany

Printed by Libri Plureos GmbH
in Hamburg, Germany